T0181442

Reinforcement Learning with Hybrid Quantum Approximation in the NISQ Context

Leonhard Kunczik

Reinforcement Learning with Hybrid Quantum Approximation in the NISQ Context

 Springer Vieweg

Leonhard Kunczik
Neubiberg, Germany

Dissertation an der Universität der Bundeswehr München, Neubiberg, 2021

ISBN 978-3-658-37615-4   ISBN 978-3-658-37616-1   (eBook)
https://doi.org/10.1007/978-3-658-37616-1

Responsible Editor: Stefanie Eggert
This Springer Vieweg imprint is published by the registered company Springer Fachmedien
Wiesbaden GmbH part of Springer Nature.
The registered company address is: Abraham-Lincoln-Str. 46, 65189 Wiesbaden, Germany

Abstract

Fakultät für Informatik
Doktor der Naturwissenschaft (Dr. rer. nat.)
Reinforcement Learning with Hybrid Quantum Approximation in the NISQ
Context by Leonhard Kunczik

Complex attacker-defender scenarios are hard to solve and require enormous computational power in Reinforcement Learning. The complexity arises from their inherent characteristics: enormous state, large action spaces, and sparse rewards. The upcoming field of Quantum Computing, with its computational properties: superposition, and entanglement, is a promising path for solving computationally complicated problems. Therefore, this work studies the combination of both in the setting of policy gradient methods in Reinforcement Learning. It further proposes a novel quantum REINFORCE algorithm based on the classical REINFORCE algorithm by utilizing Quantum Variational Circuits, a quantum hybrid approximator, to represent the policy function. The new algorithm is compared against its classical counterpart and the standard Q-learning algorithm in terms of the convergence speed and memory usage on several attacker-defender scenarios with increasing complexity. Furthermore, the algorithm is evaluated on IBM's quantum computers to examine its applicability on today's noisy intermediate-scale quantum (NISQ) hardware. Finally, an outlook is given on possible future steps in Quantum Reinforcement Learning for complex Attacker-Defender Scenarios.

Gutachter:
Prof. Dr. Stefan Pickl
Prof. Dr. Udo Helmbrecht

Acknowledgements

I would like to express my most profound appreciation to Prof. Dr. Stefan Pickl for his trust in accepting me as his student and to welcome me into his research group COMTESSA at the Universität der Bundeswehr München. I would like to thank him further for his guidance during my Ph.D., the numerous discussions, and the valuable lessons he taught me. Without his support, I would not have entered this path and finally accomplished my Ph.D.

Furthermore, I would like to thank Prof. Dr. Udo Helmbrecht for his support and trust during my research. Our long discussions and his questions that challenged my understanding and preliminary results have been crucial for this work. He encouraged me to dig deeper in uncharted territory in Quantum Reinforcement Learning, which finally led to this thesis.

I would like to especially thank Junior Prof. Dr. Maximilian Moll for his guidance as a Postdoc and his continuing support when he started his professorship. His deep understanding of Reinforcement Learning and his unique ability to comprehend new ideas greatly supported my research.

I would also like to thank my Ph.D. commission, Prof. Dr.-Ing. Markus Siegle, Prof. Dr. Uwe Borghoff, Prof. Dr. Gunnar Teege and Prof. Dr. Arno Wacker.

Finally, I would like to thank the whole COMTESSA group and my colleagues. I would like to especially mention Dr. Sorin Nistor, Dr. Son Pham, Gonzalo Barbeito, and Capt Jacob Ehrlich, whom I am thankful to be colleagues and friends.

Contents

List of Figures

List of Tables

Motivation: Complex Attacker-Defender Scenarios—The Eternal Conflict

Ultimate excellence lies not in winning every battle, but in defeating the enemy without ever fighting.

Sun Tzu, The Art of War

The story of conflicts on the earth dates back to the first living creatures on the planet. Following Darwins' theory of evolution, every living being is in constant competition for the limited resources it depends on [Darwin, 1859]. The ones best adapted to their environment survive and become partially immortal by passing on their ancestral estate to their offspring. This everlasting conflict leads to the diversity in living organisms that inhabit the earth.

When the first predators evolved that feed on other organisms, this conflict transitioned to another level. The new predators started to hunt for prey, and the first attacking scenarios appeared. Eventually, the prey started to save itself from extinction by developing a mechanism to defend themselves. This response evened out the predator's advantages, and since then, both sides are in a race to evolve a skill that tips the scale in their favor. This development marks the beginning of attacker-defender scenarios on earth [Bengtson, 2002].

Until the emergence of the homo sapiens, this gradual evolution of more advanced skills to dominate the other side, be it from the attacker or defender perspective, was driven by chance and evolution theory. However, the homo sapiens brought

Supplementary Information The online version contains supplementary material available at (https://doi.org/10.1007/978-3-658-37616-1_1).

a new aspect into this situation. The homo sapiens, as an omnivore (eating both animals and vegetables), was participating on both sides of the surrounding conflict. Nevertheless, its increasing cognitive capacity provided the ability to outperform evolution by *learning* to become a better attacker on the one hand and a more skillful defender on the other hand to secure its survival [Scull, 1992].

This advancement brought a new aspect into attacker-defender scenarios since the odds were not anymore determined by nature but instead by *human cognition* and its ability to learn from *observation*. Finally, this leads to more advanced strategies and inventing new tools for everyday use and specialized tasks. However, the homo sapiens did not only use their improving skills to survive from the everlasting thread imposed by the wild surrounding it. If too many humans gathered in one place, they competed for the limited food and water supply and save places to rest. This lead to conflicts between tribes, and the first wars between human beings appeared [Keeley, 1996].

This type of conflict revealed a new characteristic in attacker-defender scenarios since species from the same race deliberately fought against one another. As it is known from history, humanity developed even more advanced fighting tools and war techniques, culminating in World War II and others. However, one of the first records of strategic thinking in (armed) conflicts can be dated back to the Chinese general Sun Tzu, who is attributed to be the author of the Art of War written around 475 B.C.E. [Zi et al., 2009]

The Art of War has influenced military strategies for centuries. However, it is also studied in other fields like economics and politics since it provides insights into the nature of conflicts and is not restricted to warfare. Although it defined how people think about conflicts for more than 2500 years, it is also an evidence and a guideline for strategic thinking in attacker-defender scenarios in written form. It documents the beginning of a new way to think about conflicts between two parties and archives and shares this knowledge in written form with other generals, officials, and strategists [Zi et al., 2009].

The Art of War analyzes the most general form of two-party conflicts. Both parties interact with each other in an active act of warfare where one takes the hostile, attacking side and the other the defending part. Both sides follow an active strategy. At each point, they can actively decide which action they will take next to achieve their goal. This possibility could even mean that both sides can actively change their strategies or use deceptive moves to confuse and mislead the other party. This further implies that both observe the other and decide on the next step based on their current information.

However, there is another category of attacker-defender scenarios. While both parties actively take place in the conflict in the previous case, there are different situations in which only one side can actively choose its actions, and the other has to stay with a fixed strategy. The following provides an example for such a situation:

On August 21, 1911, three men hurried out of the Louvre in the center of Paris. The three carried an about 80 × 55 centimeter large rectangular object covered with a blanket and were headed to the train station to board the next train. They left the city and returned to their homes in Italy. Although it might have looked odd, three men leaving the most significant art museum in France that holds the nation's art treasures with a blanket under the arm, the men could leave the country unrecognized.

It took 28 more hours to recognize the empty location in one of the many galleries in the Louvre and to realize the missing painting. The painting stemmed out of Leonardo da Vinci's collection. He painted it in 1507, and because of its theft, it is now one of his most celebrated pieces. The three men from Italy were able to leave the Louvre with nothing else than the Mona Lisa. After 28 months of searching for the lost masterpiece and global accusations against Americans or even the German Kaiser, the Mona Lisa was found by chance in Florence.

It turned out that the three men who stole the painting were Vincenzo and Michele Lancelotti, and Vincenzo Perugia. Perugia was finally caught when he tried to sell it to an art dealer in Florence. The buyer recognized it as the original, and instead of going to the meeting, he sent the police to capture the thief. As it later turned out, Perugia worked in the Louvre to install protective measures. On Sunday night, he and his accomplices hid in one of the museum's many cabinets. When they had the chance, they obtained the painting and left in the morning. [Hoobler and Hoobler, 2009]

The above illustrates a scenario with a *fixed defense scheme*, partially or even entirely known by an *active attacker*. The static defender is characterized by the fixed security measures in the Louvre that are mainly linked to the building's structure and maybe some additional security guards. There are similar situations where the attackers' strategy is known, and the defending side has to come up with a plan for a successful defense. All three attacker-defender scenarios are interesting to study. However, this work primarily focuses on scenarios with an active attacker and a static defender. Such situations are primarily of interest to the security community, which focuses on developing strategies to prevent attacks on large targets. As in the above example, they have to plan security measures to ensure that an adverse event will not happen, e.g., the theft of Mona Lisa or a bombing at an airport or airplane [Brown et al., 2006, Alderson et al., 2011].

There are numerous ways to find an optimal strategy within such settings. The naive approach is to deduce a successful strategy from pure logical thinking by taking all possible actions and situations into account. Probably, this is what Perugia had done. He might have realized that Louvres' security system has a gap, and he came up with an idea of how to leverage on this weak spot. However, if a situation is more complex than the one above, it becomes intractable to think through all possibilities [Moll and Kunczik, 2019].

1.1   Reinforcement Learning and Attacker-Defender Scenarios

A different approach, which can handle higher complexity, to solving complex attacker-defender problems is *reinforcement learning* (RL) [Moll and Kunczik, 2019]. RL is one of the three major research streams in machine learning. It primarily focuses on complex problems that require interacting with an (unknown) environment and take place over multiple time-steps. RL can be used to derive an optimal strategy in such situations. To this end, it relies on a simulation of the problem to find and improve a strategy based on trial and error learning until an optimal strategy is obtained [Sutton and Barto, 2018].

The origins of reinforcement learning date back to the work of Edward Thorndike in 1911. In his theory *Law of Effect*, Thorndike summarized learning by trial-and-error, which he observed while studying animals' behavior [Thorndike, 1911]. In his trials, he observed that if an animal has to decide while repeatedly facing the same situation, it is more likely to take the action that leads to a pleasant situation (positive reward) while avoiding actions that end in discomfort (negative reward).

Similar to the Law of Effect, Pavlov developed his famous experiment to *reinforce* the connection between a behavioral pattern and a stimulus [Pavlov, 1927]. This was the first time that the term *reinforcement* was used in connection with learning and thus prepared the ground for RL as it is known today.

Those ideas from psychology and learning theory inspired computer scientists and mathematicians to build machines that can mimic a human brain and learn to solve tasks. Alan Turing was one of the first computer scientists who published the idea of implementing a trial-and-error-driven learning system algorithmically in 1948 [Copeland, 2004]. He was succeeded by generations of researchers that tackled the problem of trial-and-error-based learning. They developed numerous algorithms and mechanical systems that successfully solved some specific problems, like BOXES that learned to balance a pole attached to a cart (the cart pole problem) [Michie and Chambers, 1968]. While the cart pole problem is not attacker-defender specific, the idea that steered its development was to find an algorithm to play chess. However, at that time, computational resources restricted the research to more superficial problems.

A different stream of research that influenced the development of RL as it is known today, is optimal control theory. Control theory studies (dynamical) systems and tries to find an optimal control (a sequence of actions that manipulate the system) to steer it into a favorable state. Although the idea of optimal control theory has already been studied in one form or the other for more than 300 years, it became popular when enough computational resources were available, and Americans and Russians started the first rockets and satellites into space. Launching a rocket into space demanded calculating the control commands for the engines beforehand to ensure a safe flight [Sargent, 2000]. Richard Bellman dominated this research field with his theory of Dynamic Programming [Bellman, 1957]. Dynamic Programming is still considered a fundamental technique to find an optimal solution for stochastic differential equations and dynamic systems.

The third stream, which influenced the development of RL, is temporal-difference (TD) learning. TD methods elaborate on the changes in estimations of a quantity over time. Similar to trial-and-error learning, its roots are in psychology. However, Minsky drew the connection between TD learning and artificial intelligence systems [Minsky, 1954]. This lead to the development of the actor-critic framework by Barto and Sutton [Barto et al., 1983]. This framework combines a trial-and-error-driven actor that learns based on the feedback of a TD learning critic.

A combination of all three areas formed the foundation of RL, which culminated in the publication of Chris Watkin's Q-learning algorithm [Watkins and Dayan, 1992], which marked a breakthrough in RL research. By today, it is one of the best-known and studied RL algorithms. Watkins connected the trial-and-error learning stream with optimal control by leveraging TD-learning in the Q-learning algorithm. Around the same time, Gerry Tesauro developed TD-Gammon [Tesauro, 1995]. It is the first known RL program that mastered BackGammon. Both achievements were partially able due to the increased computation capabilities of the computing systems in the 1990s, which earlier work like BOXES from [Michie and Chambers, 1968] could not leverage. Consequently, those developments and successes attracted many researchers, and RL matured to a well-established research stream in machine learning.

This development shows that reinforcement learning was not purely driven by analyzing attacker-defender scenarios; however, it is commonly applied to such scenarios. TD-Gammon was the first successful application of reinforcement learning to a complex problem at that time. However, it has been succeeded by numerous applications to even more complex situations. [Sutton and Barto, 2018] provides a detailed summary on the history of RL, while the above focuses on applications in attacker-defender scenarios.

1.1.1 Today's Challenges in Reinforcement Learning and More Complex Attacker-Defender Scenarios

Q-learning, as a tabular method, is prone to the curse of dimensionality from Dynamic Programming [Bellman, 1957, Powell, 2007], which describes the rapidly growing volume of a space when adding more dimension. Thus, even though it is a powerful method, it was long unknown how to solve problems with many possible situations (states) and actions. [Mnih et al., 2015] extended Q-learning to utiliz supervised-learning methods for neural networks by studying the Atari 2600 arcade machine's game collection. They searched for a method that could solve all games without any adaptations to the algorithm for the particular problem at hand. The Atari 2600 games provide a diverse set of problems that are hard to solve for a computer algorithm. What makes the games interesting is that while the objectives of all games are different, most of the games fall into the attacker-defender category with one static party.

For example, Space Invaders is one of those games from the Atari collection. The goal is to defend earth from an alien invasion. The defender controls a laser cannon, and its goal is to shoot all descending alien ships before reaching the earth. A picture of the game is shown in fig. 1.1. This is only one example, but all 57 games follow the same spirit since they are single-player games. The games are played against an opponent with a fixed strategy. With their new approach, Mnih et al. achieved in 29 out of the 57 games a performance similar to a human reference player or above. Remarkable is that the best result is 25 times better than the maximum high score achieved by a human player. Recently, some modifications lead to an improved algorithm named Agent57 that beat the human baseline on all Atari games [Badia et al., 2020]

However, the successes from Mnih on the Atari games marked the starting point for a quest to solve even more demanding problems, which is still ongoing. A significant achievement was reached when an RL algorithm mastered the game GO in 2017. GO is a two-person board game. It was invented in Asia about 2500 years ago and still attracts people worldwide to compete and prove their cognitive abilities. The game has straightforward rules and is played on a board with 19×19 fields. Go is known for its combinatorial complexity with about 10^{170} possible token combinations on the board.

The advances in computational power and developments in neural networks combined with an adopted RL algorithm lead to solving GO. The two algorithms AlphaGO [Silver et al., 2016] and AlphaGo Zero [Silver et al., 2017] learned the game by playing against itself and achieved human-like performance. The latter algorithm mastered the game without prior knowledge and pure self-play to finally

(a) The GO game.

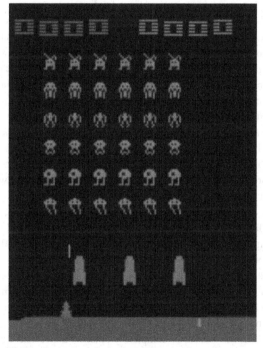

(b) The Space Invaders game from the Atari 2600 game collection [Brockman et al., 2016].

Figure 1.1 Examples of the games GO and Space Invaders

bet the GO world champion in five consecutive matches. Today, even more complex attacker-defender problems like the multiplayer-strategy game StarCraft II are about to be mastered by RL algorithms [Vinyals et al., 2019].

There is a long list of similar successes in attacker-defender situations. However, there is one more example worth mentioning that proves the power of RL in strategy development. [Baker et al., 2019] studied the child's game hide-and-seek with RL agents. In this game, one team tries to hide from the other team using different tools like movable boxes and walls to block passage for the attacking team. They analyzed different scenarios but found similar results.

Firstly, the team strategies improved alternately during the training. If one team found a winning strategy, the other team tried to counter, which finally lead to an unexpected, but superior strategy. The defending team removed all tools that the attacker needed to win and locked themselves in an unreachable place for the attacker. To come up with the final strategy would even require a human to study the problem for a prolonged time. Therefore, RL seems to be a powerful and promising tool to support the process of finding an optimal strategy in complicated and challenging decision problems, which are often the case in attacker-defender scenarios.

All the above attacker-defender scenarios in RL have one characteristic in common. One's party success can only be determined at the end of the whole chain of interactions. For example, in the game GO, the winner is only known when the game is over. Thus, during the learning process, it is only known at the end of the game whether the strategy was successful or not. Therefore, many actions have to be chosen before feedback is observed, and it is hard to conclude which specific action led to success.

While the above approaches achieved a remarkable performance, they revealed a weak spot in RL. The approaches utilize enormous computational resources and long training times. While the former problem could be solved with sufficient money, there is no obvious solution to the latter. If RL should be applied in environments that need quick solutions to changing situations, faster training is needed. One example would be the security management at an airport, where new information about a vulnerability or a new threat requires quick responses and a strategy adapted to this new information.

In RL, two different sources of complexity emerge: the classical computational complexity of the learning algorithm [Arora and Barak, 2099], and the sample complexity as it is known in machine learning [Kearns, 1990]. While the former is not commonly used in machine learning and RL, since the learning quality depends on the amount of observed data, the latter provides a tool to determine complexity in terms of the used data. This complexity definition in RL was established by [Kakade, 2003]. However, it only serves as a theoretic tool, which is not that often applied in RL. Some research is concerned with the analysis of the sample complexity of algorithms, like DSHL [Diuk et al., 2006] or TR-MAX [Oguni et al., 2014]. However, the most successes have been achieved by combining different techniques to algorithms [(Mnih et al., 2015, Silver et al., 2017, Badia et al., 2020)] and utilizing (enormous) computational power.

Therefore, complexity does refer to the problems' complexity and not the RL algorithms within this work. Thus, the complexity scales with the number of different situations which can be observed and the possible actions. An example of the problems' complexity is the enormous number of possible board positions that can occur in the game GO, which comes closer to real-world problems, like the above-mentioned airport-security example.

However, to implement such a system, new techniques are needed to either reduce the computational resources or speed up the computations. A promising solution to these problems are *quantum computers*.

1.2   Quantum Computing—An Opportunity for Reinforcement Learning in Complex Attacker-Defender Scenarios

Quantum computing revolutionized the way physicists and computer scientists think about solving challenging problems. It provides an alternative approach to computations compared to the classical binary logic concept of today's computers allows. Quantum computers are based on the physical principles of quantum mechanics to achieve computational capabilities [Scherer, 2019].

The concept of quantum computing originated from the work of physicist Paul Benioff. At the end of the 1970s, he started to work on the idea of a computer operated by quantum mechanic principles. His work culminated in a publication that, for the first time in history, brought up the idea of a quantum computer [Benioff, 1980]. Shortly after Benioffs publications, Richard Feynman caught up on this idea. He proposed that if one wants to have a precise simulation of the world, it should be built with a quantum computer [Feynman, 1982]. This initiated the development of quantum computing into the form it is known today [Nielsen and Chuang, 2011].

To understand the fundamental difference between classical and quantum computers, it suffices to compare their (smallest) information-carrying unit. In classical computers, information is stored in strings of binary *bits* and the state of a single bit is either 0/False or 1/True. The quantum computer's equivalent is a *quantum bit* (qubit) which, from a simplified perspective, can take the values 0 or 1, or any combination of them called *superposition*. Therefore, an intuitive approach to understanding the qubit is to perceive it as a random variable X following a Bernoulli distribution. Thus it is in the 1 state with probability $\mathbb{P}(X = 1) = p$ and in the 0 state with $\mathbb{P}(X = 0) = 1 - p$ during a computation [Schuld and Petruccione, 2018].

The above examples already highlight the differences between the two concepts and their implications. While a classical computer can only represent one state at a time (0 or 1), quantum computers can utilize a superposition of states, thus computing on all states simultaneously. To show the potential advantage of this property, if $n \in \mathbb{N}$ bits or qubits are used in a computation, a classical computer can compute only on one of the 2^n states at a time. In contrast, a quantum computer can use all of them simultaneously. This example already reveals why some problems can be solved faster with a quantum computer, like search in unsorted data with Grover's Algorithm [Grover, 1996].

Another concept that is not possible with classical computers is *entanglement*. This concept allows connecting qubits in such a way that their states can not be treated as independent objects, and changing the state of one will affect the other as well. This phenomenon is also known as "spooky action at a distance" as described by Albert Einstein and formalized in the Einstein-Podolsky-Rosen paradox [Rieffel and Polak, 2011]. It is a thought experiment in which entangled qubits are sent to two different faraway places. Since they are entangled, one will reveal information about the other, and therefore, information could theoretically be exchanged faster than the speed of light. This posed a conflict to the theory of relativity [Einstein et al., 1935]. The paradox was later shown to be false since the information transmission started already when the two qubits were entangled [Rieffel and Polak, 2011]. While it is essential in physics, it should serve here only as an example. The above provides some introductory background on the main concepts in quantum computing, a thorough introduction with more examples is provided in appendix A in the electronic supplementary material.

Due to the different concepts in quantum computing, it is a promising tool to solve complex problems in different disciplines like chemistry [Cheng et al., 2020, Cao et al., 2019], optimization [Farhi et al., 2014, Kandala et al., 2017] and machine learning [Schuld and Petruccione, 2018]. However, research in quantum computing is still in an early phase. Although first algorithms date back to the end of the last century, e.g., Grover search [Grover, 1996], practical experiments with quantum computers became feasible for a broader audience when IBM released their first quantum computer with five qubits to the public in 2016. Now various systems from different companies are available [Wikipedia, 2021] and IBM's largest system offers 65 qubits [IBM, 2021a]. Compared to classical computers, 65 qubits are only a tiny fraction of the bits stored and used during computation on today's hardware. For example, a classical computer needs 64-bits to represent a single double-precision floating-point number (float) for computations as specified in the IEEE 754 standard IEEE Std 754–2019, [2019]. This example provides insight into how much work still needs to be done until a full-scale quantum computer is available to perform larger computations.

Besides the small number of qubits a quantum processing unit (QPU) can handle, more challenges need to be solved to achieve the goal of a full-scale quantum computer. For example, one of the still open questions is how to realize the hardware physically. There are different approaches to create qubits. They can, for example, be implemented with trapped ions, superconductors or quantum dots. Regardless of which technology and material is used, at the moment, they all suffer from errors that distort the computations' results [Resch and Karpuzcu, 2019].

Because of the two main problems, errors and limited computational power, today's systems fall into the category of Noisy Intermediate Scale Quantum (NISQ) computers [Rieffel and Polak, 2011]. Although quantum computing is still at an early stage of development, there are already first results indicating that quantum computers will become powerful tools in the future. For example, in 2019, Google announced that they achieved quantum supremacy [Arute et al., 2019]. Quantum supremacy is the ultimate goal for quantum computers. It describes the point in time in which quantum computers can solve problems in a realistic amount of time that no classical computer can do [Preskill, 2012].

Google's claim for quantum supremacy started a considerable debate about whether supremacy was achieved or not. Finally, IBM provided a classical solution to this problem that ended the debate [IBM, 2021b], and the race for quantum supremacy is still undecided. Although RL does not seem to be a candidate to prove quantum supremacy, it is likely to benefit from the new possibilities quantum computing provides. Especially in complex problems, classical RL approaches suffer from substantial state spaces. This problem is usually solved with function approximators like deep neural networks, which reduce complex states to their core features on which a decision is based, as discussed above. Therefore, utilizing the power of quantum computers in RL seems to be a promising option to decrease the urge for massive computational power.

1.3   Specifying the Research Questions

The key question that motivates this work is whether quantum computers can help tackle RL's complexity to reduce the algorithms' computational demands. Although quantum computing is in an early stage of development, primarily due to the limited availability of real quantum hardware, its theoretical properties indicate some synergistic effects for RL. The enormous computational state spaces that already small quantum computers provide could simplify computations in the large state spaces that arise in complex attacker-defender scenarios.

To approach this question, this work is structured in the following way: Chapter 2 introduces the Information Game (IG). It is a novel attacker-defender scenario that provides the three characteristics of attacker-defender scenarios: It has a large state and action space and can be easily scaled in the state space size, and it gives only feedback if the game was won. The IG will serve as the leading example to analyze and study the possible extension of RL into the quantum domain since it can be scaled in complexity, and the optimal solution to the problem can be determined analytically.

After defining the specific attacker-defender scenario that will serve as the central object of investigation for RL methods, the necessary RL theory used throughout this work will be given in Chapter 3. The chapter first introduces the standard RL setting and mathematical notation to set the context for the various (classical) solution algorithms, which will be discussed. The chapter continues with the standard tabular Q-learning algorithm in section 3.2 to motivate the need for advanced approximation techniques in RL algorithms. Section 3.3 introduces the two standard approaches for function approximation in RL, namely value-based and policy gradient methods.

Based on the introduced standard theory of RL, chapter 4 provides a structured overview of the existing literature on quantum computing methods in RL. The overview focuses on RL methods that utilize quantum computing within their algorithms. It further identifies three different approaches to incorporate quantum computing in RL: quantum reinforcement learning (QRL), Projective simulation (PS), and Quantum Hybrid Approximation methods in classical RL methods. This structured analysis concludes that Quantum Hybrid Approximation methods have already been successfully applied in value-based RL. However, such methods have not been studied in policy gradient RL, which likely depends on sophisticated approximation methods. Therefore, the following research questions are identified:

- Can QVC be utilized to approximate the policy in policy gradient RL methods directly, and how do they perform compared to their classic counterparts?
- Can such methods be trained on today's NISQ devices without a quantum simulator, and does quantum learning provide potential benefits, like improved convergence, compared to learning with a simulator?

The former focuses on whether Quantum Hybrid Approximation methods can be used in policy gradient RL. It further aims to answer how these new algorithms perform compared to their classical counterparts. The latter question concerns whether those algorithms can already be computed with the limited capabilities of today's quantum computers.

To answer those questions, chapter 5 introduces the mathematical foundation of Quantum Variational Circuits, which is a well-established method in Quantum Hybrid Approximation [Schuld and Petruccione, 2018]. The Quantum Variational Circuits are further used in chapter 6 to define the quantum REINFORCE method. It is a novel (and the first) quantum algorithm that utilizes Quantum Hybrid Approximation in policy gradient RL, and it is the main contribution of this work to the newly emerging field of quantum RL.

The two following chapters answer the two research questions. Chapter 7 focuses on the comparison of the new quantum REINFORCE method with its classical counterpart on the different instances of the Information Game introduced in chapter 2 and chapter 8 concentrates on training the quantum REINFORCE algorithm on IBM's quantum hardware.

Based on the results of the previous chapters and the difficulties that arose while answering the research questions, chapter 9 summarizes these findings. It further provides an outlook on different directions to improve the quantum REINFORCE algorithm, especially regarding applicability to near-term quantum devices and how to utilize Quantum Hybrid Approximation in other policy gradient RL methods. This work is complemented with a conclusion in chapter 10.

The Information Game—A special Attacker-Defender Scenario

This chapter introduces the Information game (IG), a novel attacker-defender scenario. The IG studies the dynamics of an active attacker against a static defense strategy. It serves as a use case to analyze the use of quantum computing in RL and to compare a novel quantum algorithm with standard methods. Therefore, the scenario has to capture the three previously identified characteristics of attacker-defender scenarios. Thus, it should have many possible positions and actions and will only provide feedback for success.

The IG is a static attacker-defender scenario. Therefore, the defender has to commit to a static defense scheme before the attacker starts its actions. Throughout the attacker's interactions, the defender can not change his defense. As discussed in the previous chapter, these situations are common in the security community, e.g., defending a museum or airport. The scenario of the IG is motivated by the theft of the Mona Lisa in 1911. However, to better account for today's situation, the problem will be slightly modified.

With the development of the first computers and their ability to store and process vast amounts of data in the 1970s', humanity crossed over from the industrial into the new information age [Hilbert, 2020]. The industrial age was centered around the development of engines and new machines to produce goods in vast quantities in a less labor-intense way [Ashton, 1997]. The information age is defined by the possession and the ability to process large amounts of data [Hilbert, 2020]. Therefore, the values changed from physical to digital riches. To provide an example, the business from seven out of the ten largest companies by market capitalization in 2021 is based on data and information processing [Statista, 2021]. Therefore, today's most valuable commodity is information.

The above shows that while physical objects had the most value at the beginning of the 20th century, this changed in the 21st century. Therefore, the game describes the theft of information in a 2D and a 3D setting. The specific setting is as follows

L. Kunczik, *Reinforcement Learning with Hybrid Quantum Approximation in the NISQ Context*, https://doi.org/10.1007/978-3-658-37616-1_2

The IG analyzes the interaction between two competing parties. One team, the defender, tries to protect valuable information from extraction by the attacker. The attacker has a drone at its disposal equipped with sensors, e.g., audio, video, or a tactile system to gather information at a predefined position. The attacker's task is to steer the drone into the location where the information can be extracted. This position can either be where physical information is stored, like a document or hard drive, or a place from which the sensors can best receive signals without the risk of detection.

The defending team has multiple resources to prevent information extraction. They can place barriers, signal blocking devices, and surveillance appliances to stop the drone before reaching its goal. If the drone tries to pass through a barrier or to step through a wall, its position will be unchanged as illustrated in fig. 2.1; if it moves into the range of a surveillance or signal blocking device, it is detected. The defender will instantaneously destroy it as depicted in fig. 2.2.

The defender side is restricted to a fixed defense scheme. Thus, they have to decide where to place the defense systems before the game. This is similar to situations in which the security measures have to be planned in advance, e.g., when constructing a new building or planning a large event, similar to the Louvre situation.

Furthermore, the defending team has limited resources. Thus, they can not protect the whole area, such that there will always be a chance for the attacker to extract the information. However, the defenders try to reduce the chance of a successful attack with their resources by deliberately planning a defensive scheme.

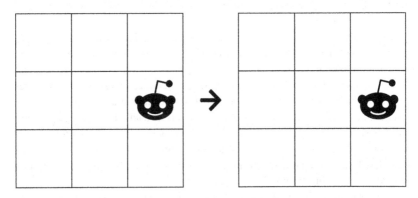

Figure 2.1 Representation of the agent moving right into a wall and not changing its position

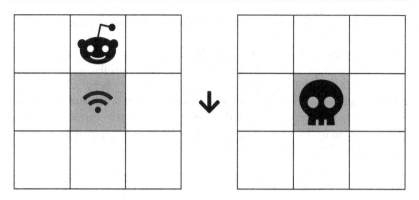

Figure 2.2 Representation of the agent moving down into a surveyed tile and being detected

As in most attacker-defender scenarios, the attacker will only observe a positive feedback if the attack was successful [Silver et al., 2017, Sutton and Barto, 2018]. This captures a crucial characteristic of attacker-defender scenarios since success is usually only known at the end. Such situations are, therefore, hard to solve because there is no feedback during the interaction that guides the solution process [Baker et al., 2019]. The attacker has to reach the goal at least once, to have some information on constructing a viable solution.

Furthermore, to make the problem scalable in complexity, it has two levels of difficulty, varying in the problem size. The first level of complexity is the 2D IG, in which the attacking party has a drone that can only move along two-dimension on a map, e.g., a driving or walking drone. High complexity is achieved by allowing defense schemes that require a flying drone that moves through a 3D space. The problem size in both situations can be varied by the size of the specific game's map. The subsequent sections describe the layout of the 2D maps with two different map sizes and the 3D IG map used within this work.

2.1   The 2D Game

The 2D game is defined by a grid in the X-Y plane of a Cartesian coordinate system. Each tile is a position the attacker can move to. The positions are enumerated from left to right, starting in the first row. The end of the grid represents a wall that can not be passed. If the attacker steps against a wall, it will not change its position. The attacker can move in each direction (left, down, right, up).

In the simplest form of the 2D game, the map is a 4 × 4 grid. This results in 16 positions, with the first position in the top left corner and the last position in the lower right corner. In this simple instance, the defense scheme relies on two cameras (depicted by camera symbols) to monitor one grid cell and one signal jammer (represented by spreading WiFi symbol) covering two adjacent cells. The attacker starts in the left upper corner, and the goal is on the opposite side of the map. The defending strategy is shown in fig. 2.3.

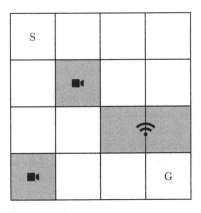

Figure 2.3 Layout of the 4 × 4 simple IG map with 4 monitored cells that lead to detection

This simple instance will serve as a starting point for analyzing the new quantum policy gradient method in chapter 7 and is referred to as the 2D4 game. Another 2D setup with increase grid size serves as a scenario with increased complexity. This instance is played on a 8 × 8 grid and is named the 2D8 game.

The 2D8 instance follows the same enumeration scheme as the smaller version, and starting and endpoint are still the first, respectively last, positions on the map. In this situation, the defender could place one surveillance camera (depicted by a camera symbol) that monitors two adjacent tiles, two signal blocking devices (represented by spreading WiFi symbol) with the ranges of 2 × 1 and 2 × 2 cells, and two motion detection sensors (visualized by a hazard symbol) covering two and three adjacent tiles. The specific defending strategy is displayed in fig. 2.4.

The 2D scenarios are inspired by the Gridworld, and the Frozen Lake problem in RL [Sutton and Barto, 2018]. In the Gridworld problem, a robot must learn to

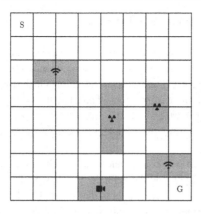

Figure 2.4 Layout of the 8 × 8 IG map with 11 monitored cells that lead to detection

navigate from a predefined starting position to a goal on a 2D grid on which it can move in all four directions. The Gridworld problem is, therefore, a 2D navigation task. The usual setting is either to reward reaching the goal or negative reward in each position except the goal. The latter provides more information for solving the problem from the RL perspective since each state provides feedback. Nevertheless, the former reflects the attacker-defender scenarios setting, where only a successful attack reveals information.

The Frozen Lake problem is similar to the Gridworld, a navigational task performed on a 2D grid. However, it further provides additional characteristics that are closer to the IG. The task in the Frozen Lake problem is to navigate from a starting position to a goal position over a partially frozen lake, which its name originates from. Since the lake is partially frozen, some position on the ice will break when stepped on and lead to drowning in the cool water. Thus, the problem is more challenging than the GridWorld problem because the weak spots on the ice are not known beforehand. The usual reward scheme is a positive reward for reaching the goal and a negative reward for falling into the lake.

The Frozen Lake problem is similar to the IG. Suppose the weak spots in the ice are interpreted as positions where the drone can be detected and is destroyed. However, the IG provides less information to solve the problem since only a success reveals feedback. The main difference is that the Frozen Lake problem is limited to a 2D layout, which the IG is not. This characteristic is unique to the IG, and the author has no knowledge that this has been done before in the literature. The details for the 3D scenarios will be given in the following.

2.2   High Complexity 3D Game

The complexity of the 2D standard situation is increased by adding a third dimension to the field. This increases the attackers' winning chance since the defender has to cover more ground to protect their secret information. On the other hand, the attacker needs to move through three dimensions. Thus, the attacker has to choose from six actions instead of four. Furthermore, the drone has to explore more space in order to achieve its goal. Therefore, the complexity increases for both parties to keep the game fair.

Within this work, an $6 \times 6 \times 6$ field is used. The attacker starts in the upper left corner of the first layer, and the defender protects the lower right corner of the last layer. The defender has two cameras that cover 4×4 and four cameras that cover 2×2 tiles to defend its secret information. The defending strategy is shown in fig. 2.5, and it is referred to as 3D6 scenario throughout the subsequent chapters.

As discussed in the previous chapter, the complexity in RL stems from the problem size, i.e., the number of positions and actions. In the 2D case, the IG scales quadratic with the map size, while the actions are fixed. The benefit of the IGs' 3D scenario is that the position space increases cubed in the map size, and the action space is increased by two actions. While the IG can be easily scaled to complex problems, in smaller scenarios, it still allows to theoretically analyze the optimal solution, which helps to compare different solution strategies.

This chapter introduced a new attacker-defender problem that can be studied with RL, namely the Information Game. Furthermore, three specific problem instances (2D4, 2D8, 3D6) have been defined, which will serve as examples throughout the work. It has been argued that the 2D situation is comparable to the famous Gridworld and Frozen Lake problem; however, it can be transferred into a 3D space, which increased the number of actions by two and drastically increases the number of possible positions. Therefore, its complexity can be scaled easily, which suits it for comparisons of RL algorithms in different levels of complexity.

The following chapter will introduce the background on RL and provides proper definitions and standard solution algorithms. It will lay the foundation for the future steps to utilize quantum computing in RL in the subsequent chapters.

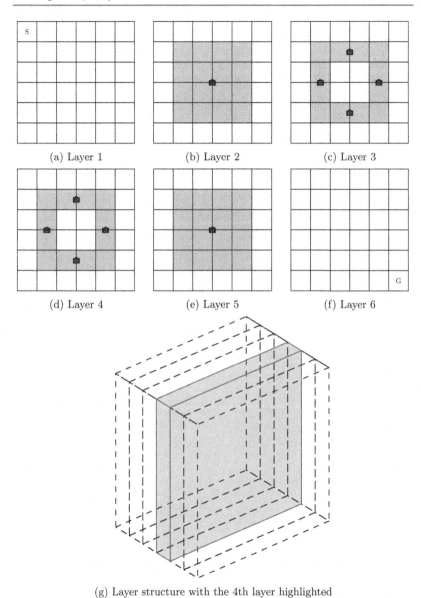

(a) Layer 1 (b) Layer 2 (c) Layer 3

(d) Layer 4 (e) Layer 5 (f) Layer 6

(g) Layer structure with the 4th layer highlighted

Figure 2.5 Layout of the $6 \times 6 \times 6$ IG map with 56 monitored cells that lead to detection

Reinforcement Learning and Bellman's Principle of Optimality

3

In chapter 1 it has been shown that attacker-defender scenarios are an important class of problems that can be found in different situations. Multiple examples have been provided, like Mona Lisas' theft in 1911, the strategic planning in warfare, or the security of large and important buildings like an airport. It has further been argued that simple methods can achieve some successes, although most attacker-defender scenarios are too hard to solve with those methods.

An alternative way to approach such problems is RL. As discussed in section 1.1.1, RL was able to solve many extremely complex attacker-defender scenarios, e.g., the game of GO, by learning from interaction with the problem (usually given by a simulation). These recent successes were possible due to two factors:

- The increase in computational power in recent time and by leveraging on expansive computational resources [Silver et al., 2017, Baker et al., 2019].
- RLs' rich theory that dates back to the beginning of the 20th century (see section 1.1).

Before turning the attention into the direction of quantum computing to answer the question of how RL can be implemented on quantum computers to reduce the demands for large computational power, to achieve similar results as in [Silver et al., 2017] in the future. This chapter will set the background on RL. As discussed earlier, RL was motivated by human trial-and-error learning. However, nowadays, it is based on a combination of different mathematical methods, like Bellman's principle of optimality [Bellman, 1957], Monte Carlo methods [Metropolis and Ulam, 1949] and temporal-difference learning [Minsky, 1954] to name just a few.

Within this chapter, the different ideas will be set into the context of RL to explain how RL methods achieve learning in a complex problem. Section 3.1 establishes a theoretical basis by providing the mathematical framework behind RL. This is

L. Kunczik, *Reinforcement Learning with Hybrid Quantum Approximation in the NISQ Context*, https://doi.org/10.1007/978-3-658-37616-1_3

further used to introduce DQN and REINFORCE two well known RL methods in section 3.3. The chapter ends with discussing the advantages and disadvantages of both techniques and their underlying concepts.

Before presenting the technical details of RL, it should be noted that the presentation in this chapter is based on the work of [Sutton and Barto, 2018] and [Wiering and van Otterlo, 2012]. For readability, references are only given if necessary. Otherwise, the two primary references are meant.

3.1   A Short Mathematical Introduction to Reinforcement Learning

In RL an *agent* tries to solve a problem, about which it has no prior knowledge on. The problem is encoded in an *environment*, with which the agent interacts with. The agent observes the *state* $S_t \in S$ of the environment at a given time point $t \in \mathbb{N}$. The variable S refers to the general definition of a state space. To give a specific example, the state space in the 2D4 Information Game is given by $S = \{(x, y) | 1 \leq x, y \leq 4, x, y \in \mathbb{N}\}$, which are all possible combinations of X-Y coordinates on the 4×4 grid.

Based on S_t the agent has to choose an *action* $A_t \in \mathcal{A}(S_t)$, which it takes in the environment. In return, the agent receives a *reward signal* $R_{t+1} \in \mathcal{R}$ and the next state of the environment $S_{t+1} \in S$. This sequence of steps is repeated, until the *episode* is terminated, $t = T_{terminal} \in \mathbb{N}$. A schematic representation of the agent-environment-interaction is depicted in fig. 3.1.

Throughout this work it is assumed that the environment's state-space S is a finite set, containing all states $S_0, S_1, \ldots S_N \in S$ with $N \in \mathbb{N}$. Each state is a unique identifier for a configuration of the environment, and it contains all the necessary information. For example, a state in backgammon comprises all black and white checkers on the board, and the state space contains all possible combinations of white and black checkers, thus all states. Similarly, the actions space $\mathcal{A}$ contains all possible actions the agent can take. It again is assumed to be finite throughout this work. However, the available actions depend on the current state. Therefore, the action space is given by $\cup_{s \in S} \mathcal{A}(s)$. In the case of backgammon, the agent can choose from a different set of actions at the beginning of the game, when all checkers are present, compared to the end of the game when only one checker is left.

The transition between states is governed by a *transition function* $T : S \times \mathcal{A} \times S \rightarrow [0, 1]$. The transition function provides for a given state $S_t \in S$ and an action $A_t \in \mathcal{A}(S_t)$ the probability to end up in the new state $S_{t+1} \in S$. This definition follows the ideas of [Wiering and van Otterlo, 2012], which deviates from the rep-

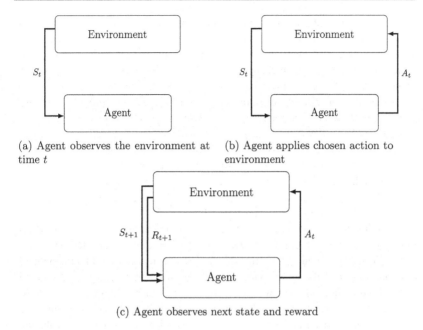

(a) Agent observes the environment at time t

(b) Agent applies chosen action to environment

(c) Agent observes next state and reward

Figure 3.1 A schematic representation of the agent-environment-interaction in RL based on [Sutton and Barto, 2018]

resentation given in [Sutton and Barto, 2018] by decoupling the reward from the transition function. For T to be a proper probability distribution, the normalization condition has to hold for each state $S_t \in S$ and action $A_t \in \mathcal{A}$ combination $\sum_{S_{t+1} \in S} T(S_t, A_t, S_{t+1}) = 1$. Furthermore, the transition function has to satisfy the Markov property, i.e., for a given history $(S_0, A_0, S_1, \ldots, S_t)$ the transition probability to the next state S_{t+1} only depends on the last state S_t and action A_t,

$$T(S_{t+1}|S_t, A_t, S_{t-1}, \ldots, S_0, A_0) = T(S_{t+1}|S_t, A_t) = T(S_t, A_t, S_{t+1}). \quad (3.1)$$

The | in the transition function indicates a conditional probability and reads as the probability to transition to S_{t+1} with the knowledge of the former state-action pair, the whole history, respectively. The triple $(S, \mathcal{A}, T)$ defines a Markov process [Cao, 2007].

The reward that the agent observes is given by a *reward function* $R: S \times \mathcal{A} \times S \to \mathbb{R}$, which assigns a numerical value to each transition. The reward defines the agents' goal in RL. However, it is the most unintuitive part of the environment. A poor choice for the reward function can increase the learning time or even lead to unexpected behavior and not the expected solution [Matignon et al., 2006]. Nevertheless, the reward is part of the environment and not the agent. It defines the problem and altering the reward function changes the problem. Therefore, it is not part of the agent's training but an external source to it.

In attacker-defender scenarios, the binary function that only provides a reward for a successful solution and no reward otherwise is a common choice for the reward function since only the final state determines the winner (see also section 1.1.1). Such situations are referred to as sparse rewards since a guiding reward is only observed in a small fraction of states. In most attacker-defender scenarios, one party wins while the other loses, and a draw is not favorable. For example, as discussed in section 1.1.1, in the game GO, the winner can only be observed once the game has ended. Thus, a reward is only given at this time.

Extending the Markov process with a reward function results in a *Markov decision process* (MDP), which is the theoretical background to study RL problems. Most algorithms and guarantees that the algorithm will *converge* (find the optimal solution) rely on the properties of the MDP [Szepesvári, 1997]. The general solution to an RL problem is a *policy*, which is a function that maps states $s \in S$ to actions $a \in A$. Thus it tells the agent which action it should take in each situation. The policy is either a deterministic function $\pi: S \rightarrow A$ or a probability distribution over actions $\pi: S \times A \rightarrow [0, 1]$ with $\sum_{a \in A} \pi(a|s) = 1 \ \forall s \in S$. The policy is often assumed to be deterministic. However, in policy gradient methods (see section 3.3.2) the policy is required to be a probability distribution over the actions.

The reward provides guidance during the solution process and the agent's objective is to find an *optimal policy* $\pi^{\star}$ that yields the highest expected *return*. The return is defined by

$$G_t = \sum_{k=t}^{T} \gamma^{k-t} R_{k+1}, \tag{3.2}$$

where $\gamma \in [0, 1]$ is the *discount factor* that determines how valuable future rewards are. For example, if $\gamma = 1$ (no discounting), all future rewards are as valuable as the current reward. The other extreme is $\gamma = 0$, and only the current reward impacts the return, while all future rewards are not taken into account. Discounting is important to ensure finding the optimal solution. For example, when considering finding the shortest path that leads to a successful attack in the 2D8 IG. If $\gamma = 1$, the action-values are binary and hence provide no information about the shortest path.

However, if, for example, $\gamma = 0.9$, the action-values are smaller for state and action pairs that are further away from the goal. Both situations are depicted in

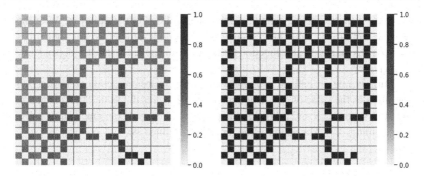

Figure 3.2 Both figures show the map of the 2D8 IG. Each state is divided into its four actions that point into the direction of travel. The left figure shows the action-value function with discounting of 0.9. The right figure shows the action-value function with no discounting $(\gamma = 1)$

fig. 3.2. The action-values for $\gamma = 0.9$ are shown on the left side. In this case, choosing the action with the highest action-value will lead to the shortest route to the goal. However, in the case with $\gamma = 1$, following the largest action-value does not help determine the shortest path to the goal. Thus, the RL algorithm can only differentiate the traveled distance to the goal based on the action-values if there is some discounting $(\gamma < 1)$. Thus in the case $\gamma = 1$ the three paths have the same value to the agent while discounting helps to find the shortest path.

3.1.1   Value Functions and Bellman's Principle of Optimality

In order to obtain the optimal policy, most RL methods rely on the concepts of optimal control theory and use a *value function* or *state-value function*. The value function estimates for each state and a given policy the expected return for visiting that state and following the given policy afterwards. Therefore, it provides a measure that says how valuable it is to be in a given state while using a certain policy.

The value function is defined by

$$v(s)_\pi = \mathbb{E}_\pi \left[G_t | S_t = s \right] \tag{3.3}$$

$$= \mathbb{E}_\pi \left[\sum_{k=t}^{T} \gamma^{k-t} R_{k+1} | S_t = s \right] \quad \forall s \in \mathcal{S}, \tag{3.4}$$

where $\mathbb{E}_\pi$ denotes the expectation value when following the policy π. With the above definition, the value function depends on a policy, therefore it is sometimes referred to as the state-value function for policy π, to explicitly state the dependency on the policy.

Similarly, one can defined an *action-value function* or *state-action-value function* that provides the value of a state-action pair. It measures the value of taking action $a \in \mathcal{A}$ in state $s \in \mathcal{S}$ and following the policy π afterwards (an practical example is given in fig. 3.3). The action-value function is defined as:

$$Q(s, a)_\pi = \mathbb{E}_\pi \left[G_t | S_t = s, A_t = a \right] \tag{3.5}$$

$$= \mathbb{E}_\pi \left[\sum_{k=t}^{T} \gamma^{k-t} R_{k+1} | S_t = s, A_t = a \right]. \tag{3.6}$$

Both value functions can be rearranged into a recursive formulation. This is essential to RL in order to utilize Dynamic Programming for obtaining a solution. The recursive relation can be seen from:

$$v(s)_\pi = \mathbb{E}_\pi \left[G_t | S_t = s \right]$$
$$= \mathbb{E}_\pi \left[R_{t+1} + \gamma G_{t+1} | S_t = s \right] \tag{3.7}$$
$$= \sum_a \pi(s|a) \sum_{s'} \sum_r T(s', r|s, a) \left[r + \gamma \mathbb{E}_\pi \left[G_{t+1} | S_{t+1} = s' \right] \right] \tag{3.8}$$
$$= \sum_a \pi(s|a) \sum_{s'} \sum_r T(s', r|s, a) \left[r + \gamma v_\pi(s') \right]. \tag{3.9}$$

The first transition in eq. (3.7) is achieved by utilizing the reward's definition from eq. (3.2). Next, the expectation value is expanded in eq. (3.8) and finally eq. (3.9) expresses the expected future reward in terms of the value function's definition from eq. (3.3). Therefore, eq. (3.9) says that the value of visiting state $s \in \mathcal{S}$ is the same as the expected reward obtained in s by following π plus the discounted value of the future state $s' \in \mathcal{S}$. This is the Bellman equation [Bellman, 1957], which is often used to solve the optimal policy.

An optimal policy $\pi^\star$ is the policy that yields the highest expected return or, more precisely, a policy is optimal if there is no other policy that leads to a larger state-value function for all states. Mathematically said $v_{\pi^\star}(s) \geq v_\pi(s)$ for all $s \in \mathcal{S}$ and it is defined as

$$v_{\pi^\star}(s) = \max_\pi v_\pi(s) \quad \forall s \in \mathcal{S}. \tag{3.10}$$

The same holds for the action-value function, and since both share the same optimal policy, they can be expressed in terms of the other. Thus, for all $s \in \mathcal{S}$ and $a \in \mathcal{A}$ the following hold:

$$
\begin{aligned}
q_{\pi^\star}(s, a) &= \max_\pi q_\pi(s, a) \\
&= \mathbb{E}[R_{t+1} + \gamma v_{\pi^\star}(S_{t+1}) | S_t = s, A_t = a]
\end{aligned}
\tag{3.11}
$$

or

$$
\begin{aligned}
v_{\pi^\star}(s) &= \max_a q_{\pi^\star}(s, a) \\
&= \max_a \mathbb{E}[R_{t+1} + \gamma v_{\pi^\star}(S_{t+1}) | S_t = s, A_t = a],
\end{aligned}
\tag{3.12}
$$

where eq. (3.12) is obtained from eq. (3.11). Furthermore, eq. (3.12) is Bellman's principle of optimality, which says that the value of a state following an optimal policy is the same as choosing the best action in the current state and then pursue the optimal policy thereafter.

There exist different techniques to solve the MDP for an optimal policy. In the classical setting considered by Bellman, the MDP (transition and reward function) is known, and the optimization problem can be solved using Dynamic Programming (DP) with backward induction [Bellman, 1957]. However, the algorithm requires the knowledge of the underlying model (MDP) and is usually computationally costly in large problems due to the *curse of dimensionality*. The curse of dimensionality describes the problem that arises in situations with many states. Since DP solves the Bellman equation recursively for each combination of states, the number of computations become soon intractable [Powell, 2007].

In the RL setting, the agent has no prior knowledge about the problem and can only observe the dynamics of the underlying MDP by interacting with the environment. The agent has to rely on observed states and rewards to solve for the optimal policy. Therefore, the agent *learns* to solve the problem since no prior knowledge exists and all information needs to be sampled from the environment. Depending on the exact method, the agent constructs approximations of the value function or other performance measures based on the observations to solve the problem. Different learning methods can be categorized by utilizing a (learned) model of the environment during the learning process. The model is used to predict the result of an action and perform some planning for future steps. These are called *model-based* RL method. The opposite are *model-free* techniques that directly learn without the planning step and therefore do not rely on a model [Wiering and van Otterlo, 2012].

Since the agent needs to sample training data from the environment through interaction, RL faces some special problems that need to be tackled in order to find the optimal solution. The first is to find appropriate approximation methods to capture enough information about the problem; the other is to balance between exploring the environment and exploiting the already observed information [Sutton and Barto, 2018]. The former will be elaborated in the next section, while the latter will be discussed next.

The problem of *exploration vs. exploitation* is well known and often discussed in RL [Sutton and Barto, 2018]. For example, in the early stage of the Information Game, when no reward has been observed yet, the agent can only explore the effect of different actions. Eventually, the agent will solve the problem by chance for the first time. In the following episodes, it has to balance between exploring other paths or following the same steps as in the successful episode to observe the reward again, which would be exploiting already acquired knowledge. Nevertheless, it is not guaranteed that the obtained solution is optimal. Therefore, the agent has to trade-off between exploring new situations and exploiting already learned paths.

In RL, the agent needs to explore the environment in the early stages of training. Thus, it will choose random actions (most of the time) until it achieves some performance level. At this point, the decision has to be made to further explore the environment to improve the current policy or to exploit the policy and securely receive the known return. One conventional approach to solve this dilemma is to use a (decaying) ϵ-greedy policy. This means that the agent selects actions greedily from the action-value function, and with some probability $\epsilon \in (0, 1)$ it takes a random action. Usually, ϵ is initialized with a large value, e.g., 0.99, and decays over time to have more exploration at the beginning and exploitation at the end.

3.2   From Bellman to Q-learning—The Tabular Approach

As already mentioned above, RL relies on different techniques to approximate (learn) a function from data, which is used to derive a policy. If the problem's exact dynamics would be known, the optimal policy could be obtained using DP methods. However, only observations (self-generated data) from interactions with the environment are available. Based on this data, the optimal policy has to be learned. One of the early methods in RL, which is still very common, is Q-learning.

Opposed to other methods, it does not rely on advanced function approximation methods but rather on the Bellman equation and bootstrapping.

Q-learning was invented by Chris Watkins during his Ph.D. research [Watkins and Dayan, 1992]. Its name stems from the fact that it learns the action-value or Q function and derives a policy based on it. Q-learning utilizes *temporal differences*, meaning that it immediately updates the action-value function after performing an action by weighting between the observed reward from the action and an estimate of the new state's value. This method revolutionized RL at that time since algorithms were usually based on Monte-Carlo approaches.

Monte Carlo methods have to go through a whole episode to compute the value of each state (and action) within the episode. Therefore, updates only take place at the end of each episode. However, bootstrapping with temporal difference learning allows direct updates to the action-values based on the estimated TD error. Another advantage of Q-learning was that it is trained *off-policy*, meaning that the approximated values of the action-value function do not depend on the policy used to sample from the environment.

Q-learning provided a method to learn from the observed state-action-reward triple directly and thus reduced required memory since the whole episode does not have to be stored, which has been an issue previously. However, state-action values are stored in a table with one entry for each state-action pair. This restricts Q-learning to problems with a moderate number of states and actions since otherwise, the Q-table would consume too much memory.

The update of the action-value function is derived from the idea of Bellman's principle of optimality in eq. (3.11) and utilizes eq. (3.12). The update-formula is given by

$$Q(S_t, A_t) \leftarrow (1 - \alpha)Q(S_t, A_t) + \alpha \underbrace{\left[R_{t+1} + \max_a Q(S_{t+1}, a) \right]}_{\text{Target}}, \qquad (3.13)$$

where α defines a weighting between the old value and the update based on the observed reward and new state, usually referred to as target. Bellman's principle is utilized within the target, by splitting the value function into the observed value and value of the new state. Thus, the approximation of the value function is the sum of the immediate reward and the next states' optimal value, derived from eq. 3.12.

Algorithm 1 Q-learning with ϵ-greedy action selection [Sutton and Barto, 2018]

Require: $Q(s, a) \in \mathbb{R} \;\; \forall s \in S, a \in A(s)$: arbitrary initial action-value function, but $Q(terminal, \cdot) = 0$

Require: $\epsilon \in (0, 1)$: Exploration parameter

Require: $\alpha \in (0, 1]$: Learning rate

Require: $N \in \mathbb{N}$: Number of iterations

  for $n = 1, \ldots, N$ **do**

    Initialize S

    while S not terminal **do**

      $A \leftarrow \begin{cases} \arg\max_a Q(S, a) & \text{with probability 1-}\epsilon \\ \text{random action} & \text{with probability } \epsilon \end{cases}$

      Take action A, observe R, S'

      $Q(S, A) \leftarrow Q(S, A) + \alpha[R + \gamma \max_a Q(S', a) - Q(S, A)]$

      $S \leftarrow S'$

    end while

  end for

The Q-learning algorithm with ϵ-greedy policy is given in algorithm 1. N determines the number of episodes for learning, while α is the *learning rate* that defines the trade-off between the old and the new value in the action-value update. The algorithm uses the usual form of the update formula, while eq. (3.13) emphasizes the underlying weighted update.

3.3   Approximation Techniques in Reinforcement Learning

The previous section introduced Q-learning, a standard method in RL. This chapter focuses on methods that use advanced function approximation techniques like neural networks to learn the value function in section 3.3.1 or the policy in section 3.3.2.

For each of these categories, one algorithm will be presented. The DQN algorithm approximates the action-value function based on subsequent states, the action that lead there, and the obtained reward similar to Q-learning. At the same time, the REINFORCE algorithm directly learns the policy from the complete history of an episode. Both rely on function approximators like neural networks or linear models to represent the action-value function or the policy.

3.3.1 DQN—Advanced Value Approximation

Q-learning can successfully solve easy problems, like navigation in a small area [Sutton and Barto, 2018]. However, as already indicated, it suffers from the curse of dimensionality, since the action-value table scales with $|\mathcal{S}| \times |\mathcal{A}|$, which becomes intractable in large state and action spaces. This makes Q-learning impractical for either too large state and action spaces or if the exact number of states is unknown beforehand. One example of such a problem is the Atari 2600 game collection [Bellemare et al., 2013]. It contains 57 arcade games, and the goal is to find an algorithm that, based on the visual state of the game (a color image with 210×180 pixels), learns to play the game while the reward is derived from the game's score.

If one tries to enumerate the state space based on the different pixel combinations, it would result in more than 28 million states. Thus, if the agent can only take four different actions, this would result in an action-value table with over 112 million entries. If each entry is considered a parameter of the action-value function, this function is almost impossible to approximate in manageable time with moderate memory usage. Therefore, instead of using a table for the action-value function, a parameterized approximation of the function can be used.

This problem seems to be appealing for a (deep) neural network or another *approximator* that is more suitable than the *tabular approach*. The neural network can be used to approximate the action-value function with less memory compared to the tabular approach. However, it is known that approximating the value function in temporal-difference learning with neural networks leads to unstable behavior [Tsitsiklis and Van Roy, 1997] and a suitable training sample with the input and the correct action-value are needed.

In 2015 Mnih et al. proposed a solution to this problem by introducing two features to the standard algorithm [Mnih et al., 2015]. First, they added a memory to the agent that stores many of the most recent state-action-reward-state sequences as a replay buffer. The replay buffer is used to randomly sample mini-batches of experiences for the neural network training since it removes temporal correlations within the data. This process is referred to as experience-replay since already gathered transitions are used multiple times for training. This is possible since Q-learning is an off-policy method. Therefore, the training is independent of the policy that was used to derive the training data.

To train the neural network, Mnih et al. use the squared difference between the current action-value and a target that is similar to Q-learning

$$\underbrace{\left(R_t + \gamma \max_a Q(S_{t+1}, a; \bar{\theta}) \right.}_{\text{Target}} - Q(S_t, A_t; \theta) \Big)^2. \tag{3.14}$$

The neural network's weights in the action-value function are depicted by θ and $\bar{\theta}$, where $\bar{\theta}$ represents the weights of a second neural network, the target network. This improvement is the second feature to increase stability during training. Suppose the target network would use the same parameters as the trained network. In that case, this could lead to large fluctuations between the current and the target values since small changes in the action-value function can lead to a different policy and, therefore, to large update values. In order to train only one neural network, the target network weights are replaced after every $C \in \mathbb{N}$ steps by θ. It has been shown that this is enough to stabilize the training.

Algorithm 2 DQN with experience replay [Mnih et al., 2015]

Require: $Q(s, a; \theta) \in \mathbb{R} \; \forall s \in S, a \in A(s)$: Action-value function with random weights θ
Require: $\bar{Q}(s, a; ; \bar{\theta}) \in \mathbb{R} \forall s \in S, a \in A(s)$: Target action-value function with initial weights $\bar{\theta} = \theta$
Require: D: Initalized replay memory with capacity $D_N \in \mathbb{N}$
Require: $\epsilon \in (0, 1)$: Exploration parameter
Require: $\alpha \in (0, 1]$: Learning rate
Require: $\gamma \in (0, 1]$: Discounting factor
Require: $N \in \mathbb{N}$: Number of iterations
  for $n = 1, \dots, N$ **do**
    Initialize S
    while S not terminal **do**
      $A \leftarrow \begin{cases} \arg\max_a Q(S, a; \theta) & \text{with probability } 1\text{-}\epsilon \\ \text{random action} & \text{with probability } \epsilon \end{cases}$
      Take action A, observe R, S'
      Store transition (S, A, R, S') in D
      Sample random mini-batch of transitions $(\tilde{S}, \tilde{A}, \tilde{R}, \tilde{S}')$ from D
      Set $\tilde{y} = \begin{cases} \tilde{R} & \text{if } \tilde{S}' \text{ is terminal} \\ \tilde{R} + \gamma \max_a \bar{Q}(\tilde{S}', a; \bar{\theta}) & \text{otherwise} \end{cases}$
      Perform gradient descent on $\left(\tilde{y} - Q(\tilde{S}, \tilde{A}; \theta) \right)^2$ with respect to θ
      Update target action-value function parameter $\bar{\theta} = \theta$ every C steps
      $S \leftarrow S'$
    end while
  end for

This approach is known as deep Q-networks (DQN), and it has led to tremendous success on the Atari games, where 29 games have been solved with comparable or better results than human players. The DQN algorithm is given in algorithm 2.

3.3.2   Policy Gradient—Policy Approximation

While DQN, as a value-based method, learns the action-value function and derives a policy from it. Policy gradient methods are fundamentally different compared to the value-based approach. Instead of learning an action-value function and then deriving a policy based on the learned values, policy gradient methods directly approximate the policy function. Usually, the policy-approximator has a *parameter* vector $\theta \in \mathbb{R}^d$ that is adjusted to fit the optimal policy. Therefore, in policy gradient methods, a policy is written as $\pi_\theta(s|a) \forall s \in \mathcal{S}, a \in \mathcal{A}$ to account for the underlying approximator and its parameters.

The policy is updated based on a *performance* measure $J(\theta)$, by performing gradient ascent on J with respect to the parameters θ. A natural selection for J is the estimated value of the starting state, following the policy $\pi_\theta(\cdot|\cdot)$. Thus,

$$J(\theta) = v_{\pi_\theta}(s_0) = \mathbb{E}_{\pi_\theta}[G_0|S_0 = s_0], \tag{3.15}$$

where $s_0 \in \mathcal{S}$ is a fixed starting state. In the case of attacker-defender scenarios the fixed starting state is not a very restrictive assumption, since for a given situation, the initial positions of attacker and defender are the same, like for example in the Information Game.

To perform gradient ascent in the form of

$$\theta_{t+1} = \theta_t + \alpha \nabla v_{\pi_\theta}(s_0), \tag{3.16}$$

where α is a given learning rate defining the updates' step-size, one needs to compute the gradient of the value function with respect to the parameter vector θ. This can be done using the *policy gradient theorem* [Sutton and Barto, 2018]

$$\nabla v_{\pi_\theta}(s_0) \propto \sum_s \mu(s) \sum_a (q_{\pi_\theta}(s, a) - b(s)) \nabla \pi_\theta(a|s), \tag{3.17}$$

where μ is the proportion of times a given state was visited following policy π_θ, $q_{\pi_\theta}(\cdot)$ is the value of the state-action pair under the given policy and $b: \mathcal{S} \rightarrow \mathbb{R}$ is an arbitrary baseline, which does not depend on a. The baseline is not necessary

since it does not change the expectation value. However, it reduces the variance of
the approximation by factoring out the dependence on all future states of $q_{\pi_\theta}(\cdot)$,
which can speed-up the learning [Sutton and Barto, 2018].

In essence, the policy gradient theorem says that the gradient is proportional to
the time spent in each state, times the value of each action weighted by the policy's
gradient. However, it is not possible to compute exact expectations for π_θ, since
the exact dynamics of the environment are not known [Sutton and Barto, 2018].
REINFORCE an algorithm presented by Williams, solves this issue by utilizing
stochastic estimates of the expectation values collected during an episode, thus
using some form of Monte Carlo simulation [Williams, 1992].

The time spent in a given state can be approximated by the state's occurrence
while following the policy. Thus, for the observed state $S_t \in \mathcal{S}$ at time t, eq. (3.17)
simplifies to

$$\nabla v_{\pi_\theta}(s_0) = \mathbb{E}_{\pi_\theta}\left[\sum_a (q_{\pi_\theta}(S_t, a) - b(S_t))\nabla\pi_\theta(a|S_t)\right]. \tag{3.18}$$

Note here, that the baseline does not change the expectation value, since

$$b(S_t)\sum_a \nabla\pi_\theta(a|S_t) = b(S_t)\nabla 1 = 0. \tag{3.19}$$

This is due to the policy's' definition, as a probability distribution over actions for
a given state.

The same approach as above can be taken for the observed action $A_t \in \mathcal{A}$ from
the policy. However, before doing this, it is convenient to write the full expectations
over actions

$$\nabla v_{\pi_\theta}(s_0) = \mathbb{E}_{\pi_\theta}\left[\sum_a \pi_\theta(a|S_t)(q_{\pi_\theta}(S_t, a) - b(S_t))\frac{\nabla\pi_\theta(a|S_t)}{\pi_\theta(a|S_t)}\right] \tag{3.20}$$

$$= \mathbb{E}_{\pi_\theta}\left[(q_{\pi_\theta}(S_t, A_t) - b(S_t))\frac{\nabla\pi_\theta(A_t|S_t)}{\pi_\theta(A_t|S_t)}\right] \tag{3.21}$$

$$= \mathbb{E}_{\pi_\theta}\left[(G_t - b(S_t))\frac{\nabla\pi_\theta(A_t|S_t)}{\pi_\theta(A_t|S_t)}\right]. \tag{3.22}$$

The first equality in eq. (3.20) derived by multiplying and dividing eq. (3.18) by
the policy. Next, the action a is replaced by the observed action A_t sampled from
π_θ in eq. (3.21). Finally, eq. (3.5) is utilized to replace $q_{\pi_\theta}(S_t, A_t)$ by the expected

reward. By the law of total expectation, this simplifies to the return, which gives eq. (3.22). Therefore, the derivative of the performance measure simply is the return minus baseline times the derivative of the natural logarithm of the policy, since $\nabla \ln(\pi_\theta) = \frac{\nabla \pi_\theta}{\pi_\theta}$. Thus, the gradient ascent update is simply

$$\theta_{t+1} = \theta_t + \alpha(G_t - b(S_t))\nabla \ln(\pi_\theta(S_t, A_t)). \qquad (3.23)$$

This formula can be easily remembered by its acronym: REward Increments = Non-negative Factor × Offset Reinforcement × Characteristic Eligibility, where the learning rate α is the non-negative factor, the return minus baseline is the offset reinforcement and the gradient represents the characteristic eligibility. In the original publications, the offset reinforcement is given as the difference between the return and a baseline $b\colon S \to \mathbb{R}$. The baseline adds stability to the training since it reduces the variance of the weight updates. However, for the sake of simplicity, the baseline is not included in this work.

Multiple algorithms fall into the REINFORCE family. This work utilizes the standard algorithm given in algorithm 3 based on the update formula eq. (3.22), with the baseline set to 0. It is referred to as Monte Carlo REINFORCE, since it utilizes Monte Carlo like updates at the end of each episode and not like temporal difference learning after each time step.

Algorithm 3 The Monte Carlo REINFORCE algorithm [Sutton and Barto, 2018]

Require: $\pi(\cdot|\theta)$: A differentiable policy with random weights θ
Require: $\alpha \in (0, 1]$: Learning rate
Require: $\gamma \in (0, 1]$: Discounting factor
Require: $N \in \mathbb{N}$: Number of iterations
  for $n = 1, \ldots, N$ **do**
    Generate an episode $\{S_0, A_0, R_1, \ldots, S_{T-1}, A_{T-1}, R_T\}$ with the current policy
    for $t = 0, \ldots, T - 1$ **do**
      $G \leftarrow \sum_{k=t}^{T} \gamma^{k-t-1} R_k$
      $\theta \leftarrow \theta + \alpha\gamma^t G\nabla \ln \pi_\theta(A_t|S_t)$
    end for
  end for

Finally, it is essential to note that REINFORCE algorithms do not rely on the policy's representation. They only require the policy approximator to be differentiable. This means that not only universal approximators like neural networks can be used. In some cases, already a linear model can solve the problem.

3.4   Policy vs. Value-based Methods in Attacker-Defender Scenarios

The significant difference between policy and value-based methods is the fact that the former directly approximate the policy, while the latter learns an action-value function first and, in a second step, derives a policy from it. This difference has an enormous impact on the approximation method used for the policy or action-value function. The action-value function must be able to represent more values compared to the policy. While the action-value function is, in some cases, very complex, the policy is usually relatively simple, and it has only a few values it needs to represent.

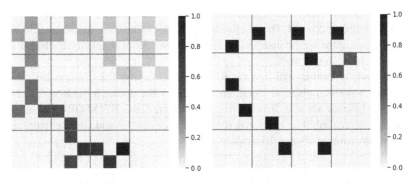

Figure 3.3 Both figures show the map of the 2D4 IG. Each state is divided into its four actions that point into the direction of travel. The left figure shows the optimal action-value function with discounting of 0.8. The right figure shows the optimal policy for each state

For example, in the 2D4 IG, the optimal policy's image is $\{0, 0.5, 1\}$, while the value functions' image contains 11 elements. The optimal value function and policy for the 2D4 IG are depicted in fig. 3.3. If the complexity grows as in the more complex 2D8 game, the policy's image does not change, while the value functions complexity grows (as depicted in fig. 3.2). Furthermore, the value function depends on the chosen discount factor; thus, the value functions' approximator essentially needs to represent the value function for all possible discount factors.

The difference in the complexity of the policy and the action-value function is, in essence, because the value function has to capture more information about the problem. However, in order to derive the policy, most information of the value function is not used. Thus, the policy has a more straightforward structure, which allows using function approximators with less *expressivity* in policy gradient methods. The expressivity correlates with the function approximators' degrees of freedoms (the number of weights). Therefore, value-based methods need more expressive approximators, like (deep) neural networks with more units and layers.

Although the above indicates that every problem can be solved with a simple approximator for the policy, some caution must be taken. For example, when the policy is straightforward, e. g., always the same action is optimal, the policy approximator still needs to represent all possible policies since they are needed for proper exploration. Thus, if the approximator is too simple, the learning can become more difficult since the agent can not explore the environment.

The beginning of this chapter introduced the mathematical foundation of RL. To achieve this, the basic agent-environment interaction as the central element in RL as shown in fig. 3.1 was described. Based on this illustration, the basic structure of both environment and agent have been defined.

The environment is described by an underlying MDP. It fully characterizes the state and action space. It further interconnects them by specifying the possible transitions between states and the actions that can achieve the transition. Furthermore, the MDP defines the reward for favorable state transitions. The fundamental element for the agents is a policy. It is a function that determines which action to chose in a given state, and the central problem in RL is to determine a policy that solves the problem defined by the MDP (or environment).

With this foundation of the basics elements in RL, the notion of the value function was introduced. The value function measures how valuable it is to be in a given state. As described in section 3.2, it can be iteratively computed using Bellman's principle of optimality and temporal-difference learning to arrive at the Q-learning algorithm. While Q-learning is a tabular method and, therefore, can only solve problems with a small state and action space, its extension to DQN was shown in section 3.3.1. DQN utilizes a function approximator to compute the value function. This allows to solve complex problems with extensive action and state spaces.

In contrast to the value-based algorithms, which first compute the value function and later derive a policy, policy gradient methods avoid this additional step and directly approximate the policy with a function approximator. Section 3.3.2 showed how the policy can be updated by utilizing the policy gradient theorem. It further introduced the REINFORCE algorithm as an example for a policy gradient method.

Finally, policy and value function for the 2D4 IG have been compared in the last section to exemplify the differences between policy and value-based methods. To summarize, this chapter provided a thorough introduction to RLs' mathematical background and introduced the two major streams. This background will provide the foundation for the next chapter in which the first step into connection RL and quantum computing are taken.

Quantum Reinforcement Learning—Connecting Reinforcement Learning and Quantum Computing

4

As already described at the end of section 1.3 and in more depth in appendix A in the electronic supplementary material, quantum computing provides multiple features that can increase computational speed compared to classical computers. Quantum methods can leverage a computational space that is exponentially large in the number of qubits. Due to quantum parallelism (superposition), all states can be used for computations simultaneously. This drives the quest to utilize quantum computing to speed up algorithms and motivates this work to study quantum computing in Reinforcement Learning (RL).

Although, the superposition principle might reduce the algorithms' training time since computations can be performed simultaneously. On today's quantum hardware, the bottleneck is loading and encoding classical data on quantum hardware, which makes it hard to use superposition for large data [Schuld and Petruccione, 2018]. However, the large computational space of quantum computers can be used for feature representation since a few qubits already span a large state space.

There is already a growing interest in quantum machine learning (QML), especially in supervised and unsupervised learning with a growing body of literature, e.g., [Schuld and Petruccione, 2018, Wittek, 2014, Wichert, 2013]. However, research on quantum methods in RL is still scarce and in an early stage. A possible explanation could be that the research community focusing on supervised and unsupervised learning methods in classical computing is already larger than for RL. Therefore, the same can be true for quantum computing.

The following provides a structured overview of the developments in quantum RL. The review focuses on RL methods where the agent relies—at least partially—

Supplementary Information The online version contains supplementary material available at (https://doi.org/10.1007/978-3-658-37616-1_4).

L. Kunczik, *Reinforcement Learning with Hybrid Quantum Approximation in the NISQ Context*, https://doi.org/10.1007/978-3-658-37616-1_4

on quantum hardware. There is a complete research track on utilizing RL for quantum state preparation and improving quantum algorithms like [Porotti et al., 2019, Niu et al., 2019, Xu et al., 2019]. Although the problem is from the quantum domain, the solution methods are still classical and therefore omitted in this work.

The review is structured into three research streams. For each stream, the underlying concept will be presented before summarizing all methods in detail and comparing them to their classical counterparts, if there is one. The result of this overview is used to further specify the overarching research question from section 1.3 and to identify a new direction for quantum computing in RL. The review is based on literature indexed by the scientific research databases Scopus and IEEE Xplore until the end of 2020 and only contains journal or conference publications.

4.1   Quantum Reinforcement Learning Methods

In 2005 Quantum Reinforcement Learning (QRL) was used for the first time in literature by [Dong et al., 2005a, Dong et al., 2005b, Chen et al., 2006]. Dong et al. proposed a novel algorithm with the name QRL, is based on classical Q-learning. Contrary to the classical implementation, QRL utilizes a policy based on a quantum actions space to obtain a natural action selection process that does not require ϵ-greedy action selection. The probability of selecting an action is encoded in the probability amplitude of the corresponding quantum state, and the action selection is implemented by measuring the superposition of possible states.

An additional step adjusts the probability amplitudes based on Q-learnings' state-action value function to update the policy. With this process, the authors propose to overcome the exploration vs. exploitation trade-off. Dong et al. utilize the amplitude amplification process together with state-action value from Q-learning to update the probability amplitude of the corresponding quantum state [Dong et al., 2006, Daoyi Dong et al., 2008].

A similar idea was carried on with a modified update formula for the probability amplitudes published under quantum-inspired Q-learning [Chen et al., 2008]. A later publication extends this concept to hierarchical Q-learning [Chen and Dong, 2012].

In a different stream of research, QRL was further extended to multiagent QRL by applying the theory of Nash Q-values [Hu and Wellman, 2003] instead of the usual Q-values [Meng et al., 2006, Tan et al., 2009].

The QRL algorithm is mostly studied in the context of autonomous navigation with an application in robot control using multiple sensors [Dong et al., 2005a, Dong et al., 2005b, Dong et al., 2006, Chen et al., 2008, Chen and Dong, 2012].

However, it was as well studied in different situations, for example, in the cognitive radio task, which is to find the best channel allocation in a wireless network to avoid frequency interference [Fakhari et al., 2013].

There are no new publications on QRL or QiQl after 2013. This might be because QRL relies on classical computers and only utilized some quantum mechanics ideas. A possible explanation is that this research stream was superseded by Projective Simulation, which can be run on quantum hardware and is a proper quantum computing method. However, since QRL is the first known RL method in the quantum domain, it should be mentioned, although it can not be computed on a quantum computer.

4.2   Projective Simulation Methods

The research on QRL was followed by Projective Simulation (PS), a novel RL framework that can be applied both on a classical and a quantum computer and therefore marks the beginning of RL in quantum computing [Briegel and De Las Cuevas, 2012]. PS is physically motivated by a model of the human deliberation and learning process, and it is less mathematically rigorous developed compared to standard methods in RL. The framework's main features are:

1. a memory,
2. a mechanism to simulate the possible implications of a decision,
3. a learning and decision method to improve performance from perceived information.

A schematic representation of the framework is given in fig. 4.1, which provides a more detailed description of the agent compared to the standard representation as shown in fig. 3.1. Therefore, PS still follows the standard agent-environment interaction in RL; however, the main difference to classical methods is that the agent is divided into different sub-components, which provide specific, physically motivated functions. Thus, PS can be seen as a new class of algorithms in RL like Q-learning or REINFORCE algorithms in the standard RL theory. The specific components in PS are: ECM, Sensors, Actuators and Learning & Decision. A detailed description of them will be given in the following paragraph.

Since PS is motivated by the human learning process, its reasoning is that an agent observes the environment with sensors, e.g., vision, touch, or distance. Based on the perception, an episodic & compositional memory (ECM) is consulted by a decision process to obtain the next action.

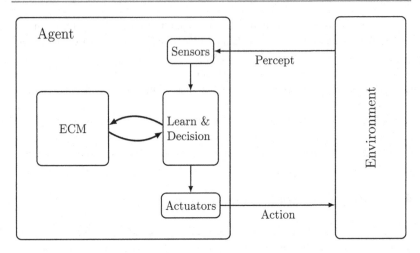

Figure 4.1 A schematic representation of the PS framework as proposed by Briegel and De Las Cuevas

Similar to the classical RL terms, an episode describes the whole chain of percepts and actions from the initial to the final state in the environment. Stored (remembered) percepts and actions are called *clips*, to take their physical meaning—a short part of an experience, e.g., a remembered voice or a sequence of images—into account. The clips can represent real or fictitious (in some way altered) experiences. Therefore, the name ECM stems from its function to store episodes as a composition of clips or even compositions of episodes.

To reflect on possible next actions and their implications for the future, the agent simulates possible futures based on the ECM. The simulation result's quality is determined by an *emotion* that is associated with an action clip. The emotions can be used to enhance the information in the ECM by realizing a short-term memory.

Once the agent has decided on an action, the action is executed by the appropriate actuator, e.g., a robot-arm, an engine, or a human hand, thus translating to an action that affects the environment. In the next step, the agent observes a new percept from the environment and a reward signal. Based on the two pieces of information, the learning method updates the ECM and the emotions to improve future actions to achieve the desired behavior.

In the original publication [Briegel and De Las Cuevas, 2012] the ECM was realized by a graph of clips, and the agent's decision and learning process was implemented by a reinforced random walk [Pemantle, 2007] on the ECM. This is the simplest way to realize a PS agent.

The basic PS method can be further extended to the reflecting projective simulation (RPS) method. In RPS, the emotions attached to an action refer to the short-term memory to remember good or bad decisions that are based on the observed reward. With the short-term memory, the agent can repeat (mixing) the random walk on the ECM until a good action (based on the emotion) is observed. Thus, the agent projects itself into different possible future situations until the deliberation process arrives at a favorable future state.

Within the publication, the authors further provided extensions to PS and RPS such that new associations can be learned within the ECM. This gives the agent the ability to perform new actions and adapt to unknown percepts; thus, it experiences some creativity and generalization to solve problems. These features are realized by combining clips and change transition weights within the graph of the ECM.

The extension from PS to RPS in [Briegel and De Las Cuevas, 2012] allows implementing the PS-algorithm on a quantum computer using a Grover-Search like approach. The basic idea is to use Grover-Search in the ECM to find good actions (marked with a positive emotion). A detailed extension of RPS to QRPS is given by Dunjko et al. by examining the implementation of QRPS on a trapped ion quantum computer [Dunjko et al., 2015].

In theory, it is proven that the random walk in quantum PS (QPS) and quantum reflective PS (QRPS) provides a quadratic speed-up in deliberation time to the classic PS algorithm. This speed-up can provide an advantage to a QPS agent in an active learning setting [Paparo et al., 2014]. The theoretical result was experimentally confirmed by analyzing the mixing time of the underlying Markov chain in QRPS, running on a trapped-ion quantum processor [Sriarunothai et al., 2019].

[Clausen and Briegel, 2018] provides an extension to QPS to incorporating eligibility traces and edge glow. The algorithm encodes perceptual inputs as a quantum state, and unitary operations (the agent's policy) are learned to map the percept to an action. The unitary operation is updated by a process similar to backpropagation in neural networks, where the observed reward gives the updated weight.

4.3   Quantum Hybrid Approximation Methods

Opposite to QRL or QRPS, the stream of quantum hybrid approximation methods (QHAM) relies on classical RL methods like Q-learning; however, instead of quantifying the agent, quantum methods are utilized to approximate a value function. Within QHAM, two research streams have emerged. One focuses on quantum annealing-based methods (Boltzmann machines), the other on quantum variational circuits (QVC).

Crawford et al. proposed to use Quantum Boltzmann Machines to approximate the Q-function in temporal difference manor, based on the initial idea of Sallans and Hinton [Sallans and Hinton, 2004]. They first extended Sallans and Hinton's algorithm to Deep Boltzmann machines by computing the system's energy by simulated (quantum) annealing [Crosson and Harrow, 2016]. In a second step, the Boltzmann machine was quantized to a Quantum Boltzmann machine, where simulated quantum annealing (SQA) is required [Crawford et al., 2018]. In a recent study, Crawfords' SQA method was extended to multi-agent RL [Neumann et al., 2020].

The QVC based QHAM methods were introduced by Chen et al. under the name Deep Quantum Learning (DQL) by replacing the neural network function approximator of DQN [Mnih et al., 2015] by a QVC (for details on QVC see chapter 5). Chen showed that the advanced machinery of QVC could be used in deep RL to learn the Q-function with less trainable parameters compared to the classical DQN by Mnih et al. They provided proof of concept by successfully solving the frozen-lake and the cognitive-radio problem. Although their algorithms were trained on a quantum computer simulator, they still could show that evaluating the trained parameters on an IBM quantum computer yields the same result as the simulator. This indicates that QVC based RL could be robust against error on NISQ hardware [Chen et al., 2020].

Lockwood and Si extended the DQL concept to double Q-learning [van Hasselt et al., 2015] by introducing a second QVC for the target network. They compare their new approach to the classical neural network-based method. Their results show the comparable performance of the quantum hybrid and the classical method. [Lockwood and Si, 2020]

4.4   Defining the Research Questions

The literature overview included QRL for the sake of completeness and because it is based on the idea of Grover's algorithm, similar to PS. QRL is a purely classical method, but it introduced the notion of quantization to RL for the first time and is therefore worth mentioning.

Historically the next step towards RL on a quantum computer was PS. Although it is driven by the same idea to use Grover's search, it is not rooted in the classical RL theory. It utilizes Grover's algorithm, a fast quantum search algorithm, to find favorable actions for a given state. Although it is a pure quantum algorithm, it has some characteristics that make it hard to implement on today's quantum hardware.

To only point out the most significant issue, two copies of the whole EMC need to be stored on the quantum computer and altered multiple times, which is not applicable on near-term quantum systems as mentioned by [Sriarunothai et al., 2019].

Therefore, QHAM seems to be the favored approach to RL for NISQ devices in the near future. However, only DQL RL has been successfully applied to existing quantum computers. Crawford et al. never left the stage of theoretical research since they did not perform their experiments on real quantum hardware.

In an unpublished paper, Levit et al. compared Quantum Boltzmann machine-based RL with SQA on a classical computer, and quantum annealing on a D-Wave system [Levit et al., 2017]. The results indicate a similar performance of the two approaches. However, the authors argue that the D-Wave hardware limits the performance of quantum annealing, which depends on the internal wiring of the Quantum Computer and the control that can be exercised on the hardware. These results could explain why Quantum Boltzmann machine-based RL has not been pursued afterward.

On the other hand, QVC already has a rich body of literature providing a decent understanding of the method and multiple theoretical results [Schuld et al., 2021, Schuld and Killoran, 2019, Suzuki et al., 2020]. They seem to be a promising candidate for function approximation on quantum systems and have already been successfully applied on quantum hardware [Abbas et al., 2021]. Therefore, QVC seems to be a promising path to connect classical RL methods with the power of QC.

The research on QVC in RL focus on action-value-based RL, more precisely DQN with the neural network replaced by a QVC. However, there are no known quantum policy-gadient RL methods. Instead of approximating action-values, QVC can be used just as well as a parameterized policy in policy gradient methods. It further provides some benefits, since a quantum state already defines a probability distribution, and therefore, it can quite naturally be extended to a policy, which is a probability distribution by definition (see section 3.1). This stands in contrast to the action-values based methods. The action-value depends on the problem formulation (environment) and can take arbitrary values which can not be directly represented on a quantum computer (see section 3.4 and chapter 5). Thus, policy gradient RL has many characteristics that can benefit from QVC.

Furthermore, as already discussed in section 3.4, policy gradient methods provide benefits in attacker-defender scenarios. Due to the simpler structure of the policy compared to the value function, the policy approximator does not need to represent as many values as the approximator for the value function. Thus, in policy based RL a simpler approximator can achieve the same results, which could reduce the training-time and the requirements on the computational power. Therefore, it seems to be a fruitful approach to extend policy gradient RL to the quantum domain.

Based on the promising results that have already been achieved with QVC in action-value based RL, the above arguments for increased performance in policy-gradient methods and the missing research on policy-based RL in QC, lead to the following two research questions:

- Can QVC be utilized to approximate the policy in policy gradient RL methods directly, and how do they perform compared to their classic counterparts?
- Can such methods be trained on today's NISQ devices without a quantum simulator, and does quantum learning provide potential benefits, like improved convergence, compared to learning with a simulator?

In order to develop a new policy gradient algorithm that utilizes quantum computing through QVC instead of a classical function approximator, the next chapter will define quantum variational circuits and provide an overview of their history and application.

Approximation in Quantum Computing   5

Through the structured literature review in chapter 6, a potential to apply Quantum Variational Circuits (QVC) in policy gradient RL was identified. QVCs' have already successfully extended to value-based RL and shown to be memory efficient. However, they have not been thoroughly compared to their classical counterparts. This lead to the two specific research questions at the end of the last chapter. Before the new REINFORCE algorithm with a QVC can be developed, this chapter will provide the details and background on QVC.

QVC where develop out of the research on machine learning methods in quantum computing. The interest in utilizing quantum computing for machine learning stems not only from the fact that the theory of both significantly relies on linear algebra. The new technology seems further promising due to the superposition of states. This allows an enormous computational space on a small quantum device that would typically require sizeable computational power on a classical computer. Therefore, quantum computing could provide machine learning models that can perform more complex tasks with fewer hardware requirements than their classical counterparts.

Therefore, the research on quantum algorithms for machine learning has become a fast-growing research field approached from multiple directions. A vital research stream for this work is quantum machine learning algorithms for function approximation. In the early phase, the primary approach was to quantize classical machine learning methods [Denil and De Freitas, 2011, Schuld et al., 2015, Gao et al., 2017] and to use quantum methods to speed up the training process of classical methods [Adachi and Henderson, 2015].

Supplementary Information The online version contains supplementary material available at (https://doi.org/10.1007/978-3-658-37616-1_5).

This early development matured into pure quantum algorithms that do not rely on quantizing classical methods. Instead, it established to use new quantum methods to achieve learning [Preskill, 2018, Biamonte et al., 2017, Verdon et al., 2018]. This development leads to the class of *quantum variational circuits*, which nowadays dominate the quantum machine learning field. This class summarizes a variety of algorithms, which are based on the principles of variational circuits and hybrid training. They are mainly used to solved problems from a different area of machine learning, namely supervised learning tasks like classification [Farhi and Neven, 2018, Suzuki et al., 2020].

Since its development, it was an open question whether they have the universal function approximation property like neural networks. This is an essential property since it proves that the QVC can approximate every continuous function, and it is needed to justify using a QVC instead of a neural network. The first results have indicated that QVCs could be used as universal function approximators in specific settings [Schuld et al., 2021]. For the general case, this was shown in a recent publication by [Goto et al., 2020], which shows that QVC is universal approximators like classical neural networks.

Before introducing the algorithm in detail, the concepts of *variational circuits* and *hybrid training* will be defined. This introduction will follow the work of [Schuld and Petruccione, 2018]

5.1   Quantum Variational Circuits—A Quantum Approximator

Variational circuits define a class of algorithms that utilize a set of parameterized gates $\{G_i(\theta_i)\}, i \in I$ with $\theta_i \in \mathbb{R}$, e.g., the rotation gate, and a set of non-parameterized gates $\{G_j\}, j \in J$ like the CNOT gate to prepare a specific goal state on the quantum computer. These two sequences construct a unitary operator

$$U(\theta) = G_{j_{d+1}} G_{i_d}(\theta_d) G_{j_d} \cdots G_{i_2}(\theta_2) G_{j_2} G_{i_1}(\theta_1) G_{j_1} \tag{5.1}$$

with the parameter vector $\theta \in \mathbb{R}^d$. An example circuit for a two-qubit system is shown in fig. 5.1 with the unitary given by

$$U(\theta) = (R_z(\theta_4) \otimes R_z(\theta_3))(R_y(\theta_2) \otimes R_y(\theta_1))CNOT. \tag{5.2}$$

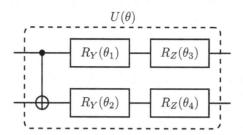

Figure 5.1 Example parameterized unitary operator on a two qubit system

The parameterized operator is applied to the zero state, thus in the above example the $|00\rangle$ state, which results in the new state $U(\theta)|00\rangle = |\phi(\theta)\rangle$. For ease of notation, in the following, $|0\rangle$ is used to describe the zero state of either a single or a multi-qubit system, depending on the situation.

This circuits parameters are trained to prepare a specific target state $|\phi_{target}\rangle$. The training, which is similar to classical neural networks, is performed on a classical computer. This introduces the second concept of hybrid training QVC since the quantum variational circuit is computed on the quantum computer, while the training that adjusts the parameters of the variational circuit is performed on a classical computer.

The training is performed by measuring the circuit and computing the error $E(\psi(\theta), \psi_{target})$, which is the difference between the states based on a specific measure, e. g., absolute error, on a classical computer. The circuit's parameters are updated using classical methods like gradient descent. This is possible since the gradient of the circuit can either be computed exactly by taking the derivative with respect to θ of the operator or numerically. This process is depicted in fig. 5.2. The concept of variational circuits was first published by [Harrow et al., 2009] as an algorithm for fast eigenvalue computation.

The quantum variational circuits' framework extends variational circuits with a data encoding, and a post-processing step [Mitarai et al., 2018]. Data encoding is necessary in order to transfer classical data to the quantum computer. There exist multiple encoding schemes like basis state encoding, where each data point in the input data is encoded as a basic state in the quantum system [Schuld and Petruccione, 2018]. For example, to encode the integer values from zero to three, two qubits are necessary, and each basis state of the quantum system is mapped to one integer value. Thus, the basis state is a binary representation of the corresponding integer. However, usually more advanced techniques are used since basis state encoding

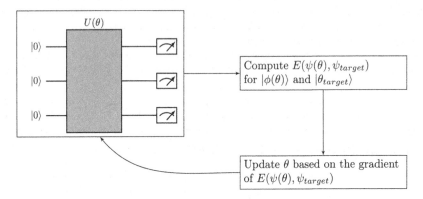

Figure 5.2 Schematic representation of the interaction between quantum and classical computer for the hybrid training of quantum variational circuits

implicitly maps each feature to a distinct position in the Hilbert space (see appendix A.1 in the electronic supplementary material), making it harder to draw a connection between similar features.

In the most general form, the features are encoded using a non-linear feature map $\phi: \mathcal{D} \to \mathbb{R}^n$ that maps features from the feature space $\mathcal{D}$ to the quantum feature vector [Schuld et al., 2020]. This feature map translates the original features to parameters of an encoding circuit U_{encode}. The encoding circuit is similar to a variational circuit, with the difference that the parameters are defined by the feature map and will not be trained.

The encoding circuit is followed by one or multiple trainable variational circuits, which can be repeated several times. The resulting unitary is

$$U(\theta, \phi(x)) = U(\theta_L)U_{encode}(\phi(x)) \cdots U(\theta_1)U_{encode}(\phi(x)), \qquad (5.3)$$

where $L \in \mathbb{N}$ is the number of repetitions of the encoding and training circuits. There are multiple ways the repetition can be implemented. A trainable circuit can be either repeated alone or together with the encoding circuit. Schuld et al. indicated that repeating both circuits can be beneficial in some cases [Schuld et al., 2021]; however, the feature map and the encoding circuit are more critical and should be chosen carefully [Suzuki et al., 2020].

With the unitary from eq. (5.3), the state of the quantum system for input x before the measurement is therefore

$$|\psi(\theta, \phi(x))\rangle = U(\theta, \phi(x))|0\rangle. \tag{5.4}$$

To measure the circuit, it is common to measure the expectation value instead of the probabilities. Therefore, the expectation value of the quantum variational circuit using eq. (A.17) in the electronic supplementary material is given by:

$$f(x; \theta) = \langle \psi(\theta, \phi(x))|\sigma_Z|\psi(\theta, \phi(x))\rangle, \tag{5.5}$$

where the observable σ_Z is chosen to be the usual Pauli Z-gate gate.

In the post-processing step, the measurement results can be adjusted to the current use case. This means that the measurements could be summed up to return one numerical value, or a bias vector $b \in \mathbb{R}^d$ could be added for prediction. Even more complex post-processing can be done by providing the results as new features to a classical linear model. In such a case, the QVC can be considered a feature extractor that extracts features from the raw data. Research indicates that for binary classification, measuring a single instead of all qubits can also provide successful results [Mari et al., 2020]. The whole quantum variational circuit framework is shown in fig. 5.3.

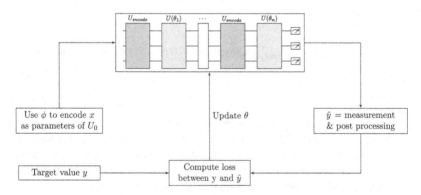

Figure 5.3 Schematic representation of the full quantum variational circuit framework

The complete training loop is similar to the training of neural networks. Based on a labeled training set $\{x_i, y_i\}, i = 1, \ldots, n$ of length $n \in \mathbb{N}$, the input feature is encoded and fed into to the quantum model. In the case where an additional bias-term is added to the measurement of the circuit during the post-processing, the prediction result is

$$\hat{y}_i = f(x_i; \theta) + b. \tag{5.6}$$

Another common approach is to use the measurement results as the features of a linear model with weights $w_j \in \mathbb{R}$ for $j = 1, \ldots, K$, where $K \in \mathbb{N}$ is the number of measured observables, which gives:

$$\hat{y}_i = \sum_{j=1}^{K} w_j f_j(x_i; \theta). \tag{5.7}$$

Independent of the chosen post-processing method, the prediction is used to compute a loss function $\mathcal{L}$, for example the mean squared error. Which is given by:

$$\mathcal{L} = \frac{1}{n} \sum_{i=1}^{n} (y_i - \hat{y}_i)^2. \tag{5.8}$$

Computing the derivative of the loss function with respect to the trainable parameters θ is possible, since $f(x; \theta)$ is a multiplication of rotation and CNOT gates. Therefore, the partial derivative of $U(\theta, \phi(x))$ with respect to a parameter is given by replacing the corresponding rotation gate with its derivative, while everything else stays the same. Thus,

$$\frac{\partial U(\theta, \phi(x))}{\partial \theta_i} = U(\theta_L) U_{encode}(\phi(x)) \cdots \frac{\partial}{\partial \theta_i} U(\theta_i) \cdots U(\theta_1) U_{encode}(\phi(x)). \tag{5.9}$$

Finally, the parameters can be updated with a suitable method like (stochastic) gradient descent or ADAM [Kingma and Ba, 2015].

The above provided a introduction to QVC with much detail on their mathematical formulation and how they can be trained with classical methods known from classical neural networks. This will be used in the following chapter, where the novel quantum REINFORCE algorithm will be developed.

Advanced Quantum Policy Approximation in Policy Gradient Reinforcement Learning

6

This chapter introduces the novel quantum REINFORCE algorithm that utilizes QVC. The quantum policy REINFORCE method relies on a similar idea to variational quantum DQL as described in section 4.3. Instead of utilizing a QVC to approximate the value function, quantum REINFORCE approximates the policy by a QVC. The policy approximator in policy gradient methods in section 3.3.2 has to be differentiable with respect to the parameters θ to use the policy gradient theorem for learning. However, as discussed in chapter 5, the gradient of a QVC can either be computed analytically or numerically. Therefore, they seem to be a good candidate to quantize policy gradient methods.

The extension of REINFORCE with a QVC to a quantum RL algorithm is given in algorithm 4. In order to quantize the classical algorithm, the policy approximator needs to be replaced with a QVC. The process is inspired by the idea to replace the neural network in DQN with a QVC as in the variational quantum DQL (VQDQL) method. [Chen et al., 2020] and [Lockwood and Si, 2020] have already demonstrated that this approach to quantizing an RL algorithm can be successful in small problems. Chen showed further that the quantized algorithm is memory efficient compared to its classical counterpart.

Quantizing the REINFORCE algorithm, as shown in algorithm 4, aims to achieve a powerful learning method that provide similar results to the classical method with less hardware and memory requirements. Smaller approximation models need less memory as well as computational resources since they depend on a reduced number of operations. Therefore, quantum REINFORCE can be a promising path to achieve a quantum RL algorithm that utilizes a quantum computer's computational capa-

---

Supplementary Information The online version contains supplementary material available at (https://doi.org/10.1007/978-3-658-37616-1_6).

L. Kunczik, *Reinforcement Learning with Hybrid Quantum Approximation in the NISQ Context*, https://doi.org/10.1007/978-3-658-37616-1_6

bilities to solve problems where the classical algorithm faces hardware limitations eventually. However, to achieve a competitive quantum REINFORCE algorithm, a considerable problem has to be overcome.

Algorithm 4 The extension of REINFORCE to a quantum policy gradient algorithm with a QVC.

Require: $\pi(\cdot|\theta)$: A QVC-policy with weights θ set to one
Require: $\gamma \in (0, 1]$: Discounting factor
Require: $N \in \mathbb{N}$: Number of iterations
  for $n = 1, \ldots, N$ **do**
    Generate an episode $\{S_0, A_0, R_1, \ldots, S_{T-1}, A_{T-1}, R_T\}$ with the current policy
    for $t = 0, \ldots, T - 1$ **do**
      $G \leftarrow \sum_{k=t}^{T} \gamma^{k-t-1} R_k$
      $\theta \leftarrow \theta + \alpha \gamma^t G \nabla \ln \pi(A_t|S_t, \theta)$
    end for
  end for

As already explored in [Moll and Kunczik, 2021], QVC in variational quantum DQL results in an extended run-time. The prolonged run-time has multiple reasons. First and foremost, it is due to the training and evaluation of the QVC. As today's research has to rely on quantum simulators instead of real hardware, quantum algorithms run slower than classical algorithms. The other reason is that depending on the simulator's implementation, the computation of the derivative of the QVC is computationally expensive.

This problem has to be solved for the quantum version of the classical REINFORCE algorithm. Fortunately, the quantum REINFORCE in attacker-defender scenarios has a considerable advantage compared to variational quantum DQL. While variational quantum DQL updates the QVC based on the replay memory after every step in the environment, REINFORCE algorithms perform the update at the end of a whole episode. Therefore, if the episode was not successful, there is no reward from which could be learned. Thus the training step can be skipped to reduce the number of derivatives that need to be computed.

This simplification does not hold for general problems. However, in attacker-defender scenarios with a binary reward scheme, this is a valid adaptation since it does not alter the general training process. A similar modification could be used in variational quantum DQL only in a more restrictive setting. If the approximator is initialized such that all Q-values are zero initially, the training could be suspended until the first success. However, it would only minimally reduce the training time of the algorithm since the Q-values will be updated after every step after the first success. The quantum REINFORCE algorithm with the new concept of conditional training is given in algorithm 5. This process is termed conditional training since it only runs through the training loop if the episode was successful. This is not a standard procedure in classical REINFORCE training. However, it is a technical trick to reduce computational time.

Algorithm 5 The quantum REINFORCE algorithm with a QVC for attacker-defender scenarios and conditional training.

Require: $\pi(\cdot|\theta)$: A QVC-policy with weights θ set to one
Require: $\gamma \in (0, 1]$: Discounting factor
Require: $N \in \mathbb{N}$: Number of iterations
  for $n = 1, \ldots, N$ **do**
    Generate an episode $\{S_0, A_0, R_1, \ldots, S_{T-1}, A_{T-1}, R_T\}$ with the current policy
    if $\sum_{k=1}^{T} R_k \geq 0$ **then**
      for $t = 0, \ldots, T - 1$ **do**
        $G \leftarrow \sum_{k=t}^{T} \gamma^{k-t-1} R_k$
        $\theta \leftarrow \theta + \alpha \gamma^t G \nabla \ln \pi(A_t|S_t, \theta)$
      end for
    end if
  end for

The conditional training alone speeds up the training process. This will become evident in the next chapter. However, to give an example, in the 2D4 IG with conditional training, the derivative is computed in only about 3% of the episodes. Nevertheless, there is one more possibility to accelerate the algorithm. At each state of the episode, the QVC has to be evaluated to obtain the policy. However, the policy does not change until a training step altered it. As argued above, the policy is only changed at the end of a successful episode. Thus, instead of evaluating the policy in every state, a dictionary can store the policy values and only evaluate the policy whenever there is no suitable entry in the storage. The resulting algorithm is shown in algorithm 6.

The stored policy could as well be used in VQDQL to some extent. However, similar to conditional training, this would only have a small effect. The action-value function is needed to derive the policy. However, after the first successful episode, due to the replay memory, the action-value function is updated with every training step. Therefore, storing the policy will not reduce the number of costly evaluations of the QVC and hence has no effect.

Algorithm 6 The quantum REINFORCE algorithm with a QVC for attacker-defender scenarios and stored policy.

Require: $\pi(\cdot|\theta)$: A QVC-policy with the weights θ set to one
Require: $\tilde{\pi}$: An empty policy-storage (e.g. dictionary)
Require: $\gamma \in (0, 1]$: Discounting factor
Require: $N \in \mathbb{N}$: Number of iterations
  for $n = 1, \ldots, N$ **do**
    Generate an episode $\{S_0, A_0, R_1, \ldots, S_{T-1}, A_{T-1}, R_T\}$ from $\tilde{\pi}$; if $\tilde{\pi}$ does not hold an
    entry for the given state, query π and store the result in $\tilde{\pi}$
    if $\sum_{k=1}^{T} R_k \geq 0$ **then**
      for $t = 0, \ldots, T-1$ **do**
        $G \leftarrow \sum_{k=t}^{T} \gamma^{k-t-1} R_k$
        $\theta \leftarrow \theta + \alpha \gamma^t G \nabla \ln \pi(A_t|S_t, \theta)$
        Empty $\tilde{\pi}$
      end for
    end if
  end for

The two changes to the quantum REINFORCE algorithm leverage on the sparse reward property of attacker-defender scenarios. Thus, they can not be generalized to every RL problem, but they counter one of the problems that make attacker-defender scenarios hard to solve. However, note that these two changes are not lmited to the quantum algorithm. They could also be implemented in the standard REINFORCE algorithm as shown in algorithm 3, if no baseline is used.

6.1   Central idea: Quantum Variational Circuits' Components

The two changes above provide the ability to speed up the costly computations of the QVC. Those changes can possibly be removed in the future once a full-scale quantum system is available for personal use, and the resources do not have to be shared with others. Until then, they provide a possibility to decrease the computational time until suitable hardware is available.

Besides the two alterations, the most significant part of the quantum REIN-FORCE algorithm is the QVC. There are many approaches how to implement them, where one has to make three decisions. The first one is to derive a suitable encoding scheme, to transfer the classical data onto the quantum system, such that the resulting quantum states can be used in the trainable layers. Next, the circuits' layout has to be determined and how often this layout needs to be repeated. Finally, the measured results have to be prepossessed to derive the final solution. The following discusses the three choices used for the quantum REINFORCE algorithm.

6.1.1   Parameter Encoding

There exist numerous ways to encode classical data in a quantum circuit. The direct approach taken by [Chen et al., 2020] and [Lockwood and Si, 2020] is to map the classical features to $[0, \pi]$ and encode the feature by a rotation. Chen et al. already argued that the process could be improved with a more advanced encoding scheme.

Suzuki et al. analyzed five different non-linear encoding functions in the context of classification [Suzuki et al., 2020]. They provided a thorough analysis of the feature maps and compared the classification accuracy on four different data sets. The results indicate that the encoding scheme in fig. 6.1 is favorable since it overall achieved good results on all four data sets. Their approach is to encode each feature on one qubit and incorporate the mixture of features further to better capture the relation of the features in the quantum domain.

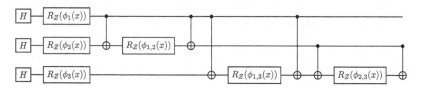

Figure 6.1 Feature encoding scheme proposed by [Suzuki et al., 2020]

The classical features are encoded on the quantum computer with the following steps. First, the qubits are initialized into the Hadamard state. Next, each feature is transferred to its qubit by a rotation around the Z-axis. The radian is given by encoding function ϕ_i, where i corresponds to the features (or qubits) index. The mixed features are again encoded as rotations around the Z-axis enclosed by CNOT gates. The rotation angle is given by the function $\phi_{i,j}$, where i indicates the index of the first feature and j is the second features index. Furthermore, j provides the qubits index the rotation gates act on. The CNOT-gates are controlled by the i-th qubit and apply their control onto the j-th qubit to entangle them.

The encoding function that achieved on average good results for all four data sets in [Suzuki et al., 2020] is

$$\phi_i(x) = x_i \tag{6.1}$$

$$\phi_{i,j}(x) = \frac{\pi}{2}(1 - x_i)(1 - x_j). \tag{6.2}$$

In Suzukis application, the feature vector x was in the range $[-1, 1]$, while in the Information Game the coordinates are mapped to the range $[0, 1]$. Therefore, $\phi_{i,j}(x)$ is altered to $\phi_{i,j} = 2\pi(1 - x_i)(1 - x_j)$ to account for the different feature ranges.

6.1.2   Variational Form

The variational form is the central part of the QVC. It defines the trainable part of the variational circuit; Thus, it determines what the QVC can learn and how many resources (qubits, parameters) are needed. Similar to the parameter encoding, there is much research at the moment focusing on the layout of the variational form. There are a vast number of approaches to implement the variational form. However, it is not known yet, which yields the best results. This problem is comparable to defining the optimal layout of a neural network. However, for QVC, there are more options one could choose from, with less understanding of what they achieve. In QVC, the variational form is built from several parameterized and non-parameterized gates.

Sim et al. studied the expressivity and entangling capacity of different variational circuits [Du et al., 2020, Sim et al., 2019]. In the context of the variational form, the expressivity measures how many different states the circuit can reach. An intuitive explanation is how much of the Bloch sphere's surface can be covered by altering the parameters of the circuit.

Sims analysis provided multiple candidates with high expressivity and entangling capacity. However, most of the circuits use a lot of gates and many parameters. This results in prolonged run-time of the circuit and is therefore not desirable for the circuit. One of the circuits they have analyzed is given in fig. 6.2.

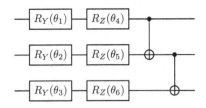

Figure 6.2 Efficient two parameter variational form analyzed by [Sim et al., 2019]

Their results show that if this circuit is at least two times repeated, it has good expressivity and entangling capacity [Sim et al., 2019]. This finding makes the circuit appealing since it has a simple structure and a manageable number of gates and parameters. For these reasons, the idea of this layout is used for the variational form in the policy QVC.

6.1.3 Post Processing

The final step after the measurement is the post-processing. The post-processing step computes the final approximation results based on the measurement outcomes of the QVC. Here, the measurements are used as features of a linear model [Schuld and Petruccione, 2018]. Hence, the policy can be derived from eq. (5.7) and is given by:

$$\pi(s, a; \theta, w) = \sum_{j=1}^{n} w_{a_j} f(s; \theta)_j, \quad \forall s \in \mathcal{S}, a \in \mathcal{A} \tag{6.3}$$

where $f(S_t; \theta)_j$ is the measured value of the j-th qubit from the QVC and w_{a_j} denotes the linear models weight for action a and qubit j. Thus, in vector notation the result can be written as the dot-product of the weight vector w_a for action a and the measurement vector $f(S_t; \theta)$

$$\pi(s, a; \theta, w) = w_a^T f(S_t; \theta). \tag{6.4}$$

The linear model is chosen because it decouples the number of qubits and the number of actions. In VQDQL, each action is represented by one qubit [Chen et al., 2020]. However, it might be beneficial for complex situations to use more qubits than actions to achieve good approximation results. Using a linear model at the end provides further an intuitive interpretation of the full approximator. The extensive computational space of the quantum computer transforms the input state into features for the linear model. Thus, the challenging task of feature extraction is performed on the quantum computer while the results are finally computed on the classical computer.

A neural network could replace the linear model to achieve the same results. The neural network's advantage is its superior approximation capability. However, the better approximation method comes at the price of decreased explainability. Furthermore, the neural network requires more computational power on the classical computer due to a larger number of parameters, contrary to the idea of transferring the computationally expansive part to the quantum computer. Therefore, a linear model is used in this work.

The process of approximating the policy with a QVC and its training is depicted in fig. 6.3. The training begins on the left side with the environment's state S_t that is encoded by applying the encoding function ϕ and provided to the QVC. The QVC is evaluated, and its result is used to compute the policy as in eq. (6.3).

Finally, the REINFORCE update rule is computed based on the chosen action A_t and the return G, which is used to update the parameters θ of the QVC and the

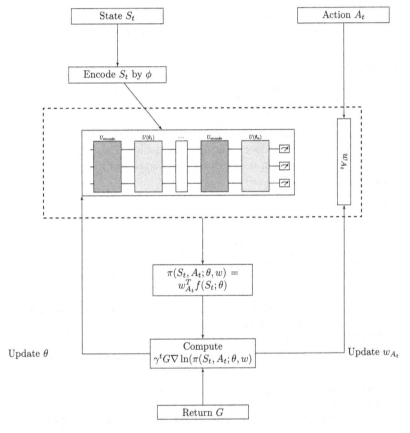

Figure 6.3 Schematic representation of the QVC training in the quantum REINFORCE algorithm

weights w_{A_t} of the linear model. With this, the training loop for one training sample (state and action pair) is concluded.

6.2   Experimental Framework & Hyper-Parameter Optimization

Before going into the details of applying the quantum REINFORCE method to the Information Game in the subsequent chapters, a few words should be spent on the implementational details. The complete code for this work is written in Python using different libraries. The source code can be found in appendix C in the electronic supplementary material. Next to the standard Python libraries like Numpy, Matplotlib, and Seaborn, the following additional libraries are used:

- *OpenAi Gym* [Brockman et al., 2016] is a standard library for reinforcement learning environments. It provides numerous predefined problems. More importantly, it defines a standard interface for environments such that the reinforcement learning algorithms can be implemented independently of the actual environment. The IG is implemented as a Gym environment, which is based on the standard Frozen Lake problem.
- *TensorFlow* [Abadi et al., 2016] is a machine learning framework specialized for fast and scaleable differentiable programming implementations. It provides an easy-to-use interface to utilize graphics processing units (GPUs). Furthermore, TensorFlow introduced a quantum machine learning implementation for quantum-hybrid machine learning. TensorFlow Quantum provides a fast quantum simulator that can be used with automatic differentiation and a larger number of qubits.
- *PennyLane* [Bergholm et al., 2018] is a machine learning library for quantum computing. It provides multiple fast and easy quantum-hybrid machine learning features, e.g., different gradient and weight-update methods. Furthermore, it can be used as an interface to implement machine learning algorithms on different quantum systems. PennyLane is used to combine TensorFlow Quantum and IBM's quantum hardware.

The Universtät der Bundeswehr München is an IBM Q-Hub and a member in the IBM quantum network. The Q-Hub provides full access to all IBM quantum systems. Therefore, the experiments with real quantum systems in chapter 8 are performed on IBM quantum computers accessed over their cloud interface called quantum experience [1].

This chapter developed the novel quantum REINFORCE algorithm, which utilizes a QVC as a function approximator for the policy. The next chapter will analyze and compare the new algorithm on the three IG instances to its classical counterpart and the standard Q-learning method. These results should answer the first research question and show whether the quantum REINFORCE is a promising candidate to reduce computational resources in complex attacker-defender scenarios.

---

[1] https://quantum-computing.ibm.com/

Applying Quantum REINFORCE to the Information Game

This chapter analyzes the results of the novel quantum REINFORCE method applied to the three instances of the IG presented in Chapter 2. The analysis starts with the basic 2D4 game, and the complexity is gradually increased until the 3D6 IG. The first research question is answered by comparing the quantum REINFORCE algorithm to classical Q-learning and REINFORCE and analyzing the differences in learning behavior and memory consumption. Since an RL algorithm's observed result heavily depends on its hyper-parameter configuration, the best parameter configuration is obtained using a hyperparameter optimization software. Before providing the results, the methodology for the experiments will be given.

7.1 Experimental Set-Up: Optimal Parameter Configuration

In RL, the agent has to produce its training data, which is significantly different from other machine learning techniques. The data generation is usually referred to as the exploration phase. Thus, to achieve reliable results in a series of experiments, the data creation process must be the same for each experiment. It should further produce the same results if it is repeated multiple times. Both ϵ-greedy and policy gradient methods rely on random numbers generated during the training. It is common practice to initialize the random number generator with a fixed seed to observe the same behavior if the experiment is repeated reliably.

Supplementary Information The online version contains supplementary material available at (https://doi.org/10.1007/978-3-658-37616-1_7).

While a fixed seed ensures reproducibility, it can bias the experiments since the algorithms' exploration depends on random numbers. Suppose the random number generator is initialized in a lucky configuration. In this case, the number of episodes until the agent reaches the goal for the first time can be very small, while in another configuration, many more episodes need to be observed for the same result. This issue is overcome by training one agent multiple times with different seeds. In this study, the agent is trained six times with different seeds to ensure that agents' configuration generalizes well.

Before continuing with the details on the hyper-parameter optimization, it is essential to note that the term agent has a slightly different meaning in this chapter than in the previous chapter. In general, in RL, an agent refers to the used training algorithm. In this chapter, an agent is understood as the used algorithm plus a specific parameter choice since the hyper-parameters define the agent's actual performance. For example, quantum REINFORCE is the learning method, while an agent further needs specific values for the learning rate (α), the discount factor (γ), and a VQC.

To find the optimal agent (parameter configuration that achieves the best performance) for each learning algorithm, the hyper-parameter optimization software Optuna is used [Akiba et al., 2019]. Optuna is a Python software that performs an efficient search through the parameter-space. This software provides the opportunity to automatically search for the optimal parameter configuration in a predefined parameter range. This process has usually been performed manually. However, automating the search benefits the comparability of the results. A fixed algorithm will ensure that the results will be the same on each run, while manually searching can be biased and lack reproducibility.

Furthermore, Optuna is very efficient in hyper-parameter optimization. It is based on an objective function that measures the agent's performance for a given set of hyper-parameters. Optuna maximizes (or minimizes) the objective function with the tree-structured Parzen Estimator (TPE) optimization algorithm [Bergstra et al., 2011, Bergstra et al., 2013]. TPE uses Bayesian updates to model the value of the objective function given the hyper-parameters. This model is used to sample new parameters to be evaluated next. This process is repeated for a fixed number of trials or until a termination criterion is reached. In this work, the number of hyper-parameter optimization steps is chosen, such that 5% of the parameter space can be explored. This specific value is chosen based on experimental results. There are no known results for the optimal number of optimization steps.

For the hyper-parameter optimization, the agent's performance is based on the number of episodes needed to converge to the optimal solution. The agent has converged if the optimal solution was obtained in a fixed number of consecutive evaluations. Therefore, the agent's performance is evaluated after each episode of training. If the stopping criterium is fulfilled, the training is ended. Otherwise, the training is terminated after a maximum number of episodes is reached.

This termination criterion can only be implemented since the optimal solution for all three IG instances is known in advance. However, it provides the advantage that the agents can be optimized in terms of their convergence speed. This provides the unique opportunity to compare the different algorithms on their convergence rates. Ideally, the algorithm should only observe a few episodes to achieve optimal performance. However, such an analysis can only be done if the optimal solution is known beforehand.

The agent's performance measure combines three factors. The first factor captures the number of episodes until the agent converged for all six seed values. This is given by

$$\sum_{i=1}^{6} \frac{n_{episodes}(i)}{max_{episodes}},$$

(7.1)

where $n_{episodes}(i)$ is the number of episodes until convergence for the agent with the i-th seed-value and $max_{episodes}$ is the maximum number of episodes at which the training is terminated.

The second part incorporates a trade-off between maximum convergence rate and the function approximators parameters, e.g., the weights in the neural network or parameters of the VQC. It penalizes large models that only marginally improve the convergence. Larger models better express complex functions, and not penalizing can lead to the behavior that always the largest model is used. The amount of parameters that define the models' memory consumption is measured as the ratio:

$$\frac{n_{param}}{max_{param}},$$

(7.2)

where n_{param} is the number of parameters of the agents model and max_{param} is the maximum number of parameters. This factor is essential in analyzing the results since the agents' memory consumption is the second comparison criterion besides the convergence rate.

To achieve a balanced trade-off between convergence and memory size, the two factors are weighted in the following way:

$$\frac{1}{7} \left(\sum_{i=1}^{6} \frac{n^{i}_{episodes}}{max_{episodes}} + \frac{n_{param}}{max_{param}} \right).$$

(7.3)

Thus, the convergence speed is emphasized more with a total $\frac{6}{7}$ of the weight placed on it, while number of parameters is less significant with the remaining $\frac{1}{7}$ of the weight.

The last factor penalizes situations in which the training exceeds the maximum number of training episodes for a seed-value. To ensure that the agent converged for all seed-values, a large penalty is given for each training that did not finish. This is measured by:

$$10 * \sum_{i=1}^{N} I_{n_{episodes}(i)=max_{episodes}}, \tag{7.4}$$

where I is the indicator function that is 1 if $n_{episodes}(i) = max_{episodes}$ and zero otherwise. The full objective function for hyper-parameter optimization is therefore

$$\frac{1}{7} \left(\sum_{i=1}^{6} \frac{n_{episodes}^{i}}{max_{episodes}} + \frac{n_{param}}{max_{param}} \right) + 10 * \sum_{i=1}^{N} I_{n_{episodes}(i)=max_{episodes}}. \tag{7.5}$$

The above objective function is specifically designed for this experiment to achieve a parameter configuration that weights the number of parameters in the approximator against convergence speed. Similar to the number of optimization runs to obtain the optimal configuration, no known literature covers the definition of an objective function that fits this purpose. Hence, the above objective function was developed within this work.

The whole experimental scheme is as follows: The above optimization criterion is used to derive the optimal parameter configuration for each learning algorithm. Afterwards, the best agent is trained again for more episodes than $max_{episodes}$. After each training episode, the agent is evaluated. The averaged evaluation results and the standard deviation are shown in the results section to compare the algorithms.

7.2   Results

The overall structure of the analysis of the results is as follows. First, the three methods are applied to the IG to determine a configuration that provides optimal results by utilizing the hyper-parameter optimization techniques of [Bergstra et al., 2011]. This is followed by comparing the convergence, memory, and parameter-setting of the three algorithms to identify differences between the quantum and the classical methods. These steps are performed for all three IGs' difficulty levels. The analysis is performed in the spirit of [Wang et al., 2019].

7.2.1 The Simple Problem—A First Approach

The 2D4 IG is used to gather first experiences with the quantum REINFORCE algorithm. Its simple structure and the small number of states provide an excellent baseline for evaluating the new algorithm and enhancing its performance. With just 16 states and six steps from start to goal, the 2D4 can be solved in a feasible amount of time with the quantum REINFORCE algorithm. It, therefore, allows analyzing different features and parameter configurations in a reasonable time.

Each of the three algorithms is optimized with Optuna on specific parameter ranges. For Q-learning the learning rate (α) is fixed on [0.2, 0.4], the discount factor (γ) is chosen from [0.8, 0.99] and ϵ is set to values in [0.9, 0.99]. The three parameter ranges are discretized with a steps-size of 0.01, leading to 4,200 possible parameter combinations. Therefore, 210 optimization runs need to be computed to search 5% of the parameter space, as discussed above.

In this setting, the optimal parameter values are not unique. The optimization runs obtained that the optimal value for ϵ is 0.96, while the learning rate and discount factor have 86 possible combinations that lead to the same result. The shown parameter in table 7.1 is only one of the 86 possible combinations. This pair is displayed since it was used to compare Q-learning with the other two methods in fig. 7.1. The whole list of parameter values is given in appendix B.2 in the electronic supplementary material.

For the classical REINFORCE algorithm, the learning rate (α) is in [0.05, 0.08] with a discretization step-size of 0.002 and the discount factor (γ) ranges from 0.8 to 0.9 with step-size 0.01. The neural network has either 1, 2, or 3 layers with the number of units in [5, 30] discretized with increments of five. The parameter space has 3,168 possible combinations, which leads to 159 hyper-parameter optimization runs.

The obtained result from Optuna for the classical REINFORCE is a unique set of optimal hyper-parameters. The best neural network configuration has one layer with five units, and the optimal learning rate and discount factor are 0.068 and 0.86.

The learning rate (α) for quantum REINFORCE is in the range [0.35, 0.55] and the discount factor (γ) lays in the interval [0.8, 0.99] for the hyper-parameter

Table 7.1 Parameter ranges for hyper-parameter optimization and optimal values for Q-learning, REINFORCE and quantum REINFORCE in the 2D4 problem

Method	Parameter	Range	Step-size	Optimum
Q-learning	Learning rate	[0.2 , 0.4]	0.01	0.26*
	Discount factor	[0.8, 0.99]	0.01	0.84*
	Epsilon	[0.9 , 0.99]	0.01	0.96
	Number of optimization runs		**210**	
			*multiple optimal values	
REINFORCE	Learning rate	[0.05 , 0.08]	0.002	0.068
	Discount factor	[0.8, 0.9]	0.01	0.86
	Number Layers	[1 , 3]	1	1
	Number Units	[5 , 30]	5	10
	Number of optimization runs		**159**	
Quantum REINFORCE	Learning rate	[0.35 , 0.55]	0.01	0.43
	Discount factor	[0.85 , 0.99]	0.01	0.95
	Number Layers	[2,4]	1	2
	Circuit Copies	[1,2]	1	2
	Number of optimization runs		**95**	

optimization. Both ranges are discretized with step-size 0.01. The layout of the QVC can take 2, 3, 4 trainable layers, and the whole circuit is at most copied twice. This results in a QVC using either 2 or 4 qubits since the circuits' input features are the X and Y coordinates on the 2D4 map.

The parameter space has in total 1,890 possible configurations, which leads to 95 optimization runs. The best configuration is a QVC with two layers and two copies combined with a learning rate of 0.43 and a discount factor of 0.95.

The training results with the optimal parameter configuration for each method are shown in fig. 7.1. Q-learning needs the least number of training iterations until all six agents have obtained the optimal solution. The first agent solves the problem after 73 iterations, while the last agent needs 310 iterations. Q-learning is followed by the quantum REINFORCE algorithm, where the first agent learns the optimal solution after 75 iterations and the last agent finishes after 838 iterations. Interestingly, the first two agents of the quantum method and Q-learning solve the problem in almost the same number of iterations. The following three agents need an additional 100 iterations to achieve the same results compared to Q-learning. However, the last

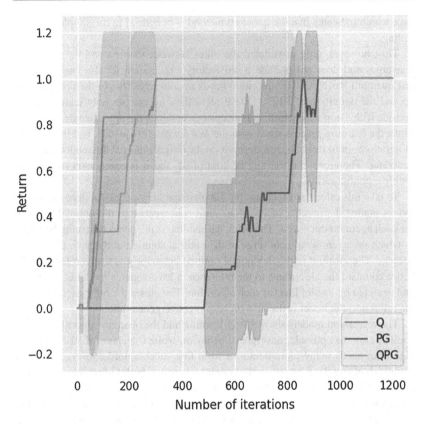

Figure 7.1 Training results for Q-learning, REINFORCE and quantum REINFORCE for the 2D4 problem

quantum REINFORCE agent needs many more training iterations to arrive at the optimal solution.

Classical REINFORCE is the slowest algorithm. The first solution is obtained after 504 iterations. This is more than 250 iterations after all six Q-learning agents learned the optimal policy. The last REINFORCE agent finished the training after 929 iterations. If only the iterations between the first and the last agents that solve the problem are measured, classical REINFORCE and Q-learning perform equally well. However, in the small problem instance, the classical REINFORCE method is slower, primarily until the optimal solution is found for the first time, than the other two, which indicates that the quantum method is beneficial in the simple problem setting.

This, however, only considered the times between the first and the last agent converges and does not provide a satisfactory explanation for the long time the last quantum REINFORCE algorithm needs to converge. Due to the problem set-up and the definition of REINFORCE algorithm, agents can only learn from an episode with a non-zero return, thus an episode in which the final state was reached. Thus, the learning process starts with the first successful episode, and the number of iterations, until this happens, depends on the initialization of the random number generator. Therefore, this can not be controlled, and it is one reason to evaluate multiple configurations.

To take this effect into account, fig. 7.2 plots the first successful iteration on the x-axis against the number of iterations until convergence to the optimal solution for each agent on the y-axis. The black dashed line is the perfect learning behavior in which the agent would converge to the optimal solution at the time of the first success. Therefore, it is the theoretical optimal behavior given as a reference. To better compare the algorithms to the reference, a least-squares model is computed and depicted by a solid line for each algorithm. The closer the regression line is to the reference, the better is the first-success to convergence ratio.

The regression models show that Q-learning and the quantum algorithm have a similar, close to optimal, convergence behavior, while the classical REINFORCE algorithm needs many more iterations from first success until convergence. This already explains the plateaus in the training results of both Q-learning and quantum REINFORCE in fig. 7.1. Both plateaus are due to late observation of the first successful episode in the initial phase of training. Figure 7.2 further explains the slower convergence of the classical REINFORCE algorithm. It needs to observe more episodes until it converges to the optimal policy.

To further analyze the convergence rate, statistics about the first successful episode until the convergence of the agents is given in table 7.2. They are computed by taking the average time between the iteration in which the first success occurred and the training converged. This factors out the random behavior until the first information from which the training starts. This further allows analyzing the average number of policy updates that the algorithm needs to observe for the two REINFORCE algorithms. For Q-learning, the number of value-function updates during the training does not depend on the number of successful episodes and is therefore omitted.

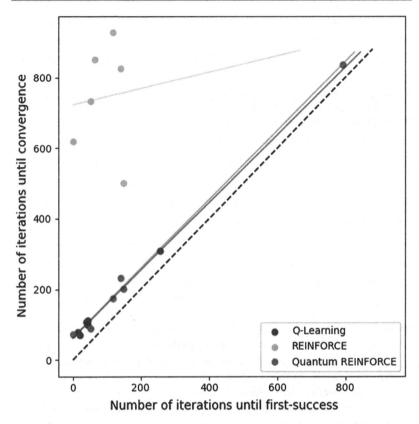

Figure 7.2 Scatter-plot of the iteration of the first-success against the iteration of convergence for the three algorithms on the 2D4 problem instance, with a least-squares regression on the data points for comparison to the theoretical optimal first-success to convergence ratio

For the first success until the convergence of the algorithm, the quantum REIN-FORCE algorithm needs, on average, the least amount of iterations and therefore shows a superior convergence speed on the 2D4 problem instance, which can not be seen from fig. 7.2. It needs, on average, 59 iterations to solve the problem after observing the first successful episode. This is slightly better than classical Q-learning, which needs 59.66 iterations on average and an enormous improvement on the classical REINFORCE algorithm, which needs one order of magnitude more iterations. However, comparing the iterations' standard deviation reveals that the same behavior between the two REINFORCE algorithms is observed. With a standard deviation of 18.15, the quantum version shows a steadier convergence rate in contrast to a standard deviation of 146.75 in the classical version.

Table 7.2 Comparing mean and standard deviation for each algorithm between the first-time reaching the target until convergence and the mean and standard deviation for the number of observed training samples for the 2D4 problem

Algorithm	First-success to convergence		Policy updates	
	Mean	Std	Mean	Std
Q-learning	59.66	7.45	–	–
REINFORCE	651.83	146.75	64.33	23.73
Quantum REINFORCE	59.00	18.15	23.83	3.48

Nevertheless, both algorithms can not compete with Q-learning. This is because policy gradient methods (like REINFORCE) suffer from increased variance compared to value-based methods like Q-learning. The fact that the quantum method's standard deviation is closer to Q-learning than its classical counterpart indicates that the quantum REINFORCE algorithm stabilizes the training.

In essence, the results in fig. 7.2 and table 7.2 show that, on average, quantum REINFORCE needs the least amount of iterations between the first success and convergence. However, its standard deviation is larger than for Q-learning. This explains the prolonged time that the last agent need until convergence in fig. 7.1. This is due to the fact that the agent does not observe the final state, as the only state from which it can learn, shortly before it converges at iteration 836. Thus, although the quantum REINFORCE method seems to converge slower than Q-learning, on average, it needs the same amount of iterations from the first success to convergence.

Comparing the number of parameters the three methods need, the quantum method is the most efficient. Quantum REINFORCE needs only 16 weights in the QVC and an additional 16 parameters for the linear model, which sum up to a total of 32 parameters. With 64 trainable parameters, Q-learning needs twice as much as the quantum method, and standard REINFORCE uses in total 72 weights. Therefore, quantum REINFORCE needs less than half the trainable parameters to compare to the classical methods. These results are similar to the statement from [Chen et al., 2020] for variational quantum DQL that RL methods with a VQC benefit from fewer trainable parameters and are therefore more memory efficient.

7.2.2   Enlarging the State Space—The 2D8 Game

The above results have shown that the quantum REINFORCE algorithm outperforms the classical REINFORCE algorithm on the simple 2D4 problem. However, to investigate if the same holds for more complex environments with a larger state space and more complex policies, the three algorithms are compared on the 2D8 IG. In contrast to the 2D4 problem, the 2D8 environment's state space is four times larger. In order to compare the three algorithms, the optimal hyper-parameter configuration has to be obtained first.

For the Q-learning algorithm, the same parameter range as in the 2D4 setting was used (see previous section). Similar to the 2D4 results, only ϵ has a unique optimal value of 0.96. The optimization algorithm obtained 78 combinations of learning rate and discount factor that lead to the same results. The learning rate of 0.25 and the discount factor of 0.85, shown in tab. 7.3, is one of the combinations, which is used for the comparison in fig. 7.3. The whole list of possible combinations is given in appendix B. in the electronic supplementary material.

The parameter range of the classical REINFORCE algorithm is adjusted for the 2D8 problem. The learning rate ranges from 0.01 to 0.03 and is discretized in 0.002 steps. The discount factor is in the range [0.9, 0.99] and discretized in steps of size 0.01. The neural network for the 2D8 environment has up to three layers with the number of units stemming from the row of five. This induces a parameter space with 3,300 possible configurations, which requires 165 parameter-optimization computations. The classical REINFORCE algorithm's optimal configuration is a neural network with one layer and 45 units and a learning rate of 0.026 together with a discount factor of 0.99.

For the quantum REINFORCE method, the learning rate is from the interval [0.25, 0.45] and the discount factor is in the range [0.85, 0.99]; both intervals are discretized with a step-size of 0.01. The QVC has 2 to 4 trainable layers, and the layout is 2 or 3 times copied, which results in a circuit on either four or six qubits. This induces a parameter space with 1,890 possible combinations, which require 95 optimization steps. The best configuration is a QVC with three trainable layers and three circuit copies combined with a learning rate of 0.3 and a discount factor of 0.87.

The training with the optimal parameter configuration for each method is visualized in fig. 7.3. The results show that in the 2D8 environment, similar to the 2D4 problem, quantum REINFORCE and Q-learning have an equally good performance. Again, classical REINFORCE needs the most time to solve the task. Analogously

Table 7.3 Parameter ranges for hyper-parameter optimization and optimal values for Q-learning, REINFORCE and quantum REINFORCE in the 2D8 problem

Method	Parameter	Range	Step-size	Optimum
Q-learning	Learning rate	[0.2 , 0.4]	0.01	0.25*
	Discount factor	[0.8, 0.99]	0.01	0.85*
	Epsilon	[0.9 , 0.99]	0.01	0.96
	Number of optimization runs		**210**	
			*multiple optimal values	
REINFORCE	Learning rate	[0.01 , 0.03]	0.002	0.026
	Discount factor	[0.9, 0.99]	0.01	0.99
	Number Layers	[1 , 3]	1	1
	Number Units	[5 , 50]	5	45
	Number of optimization runs		**165**	
quantum REINFORCE	Learning rate	[0.25 , 0.45]	0.01	0.3
	Discount factor	[0.85 , 0.99]	0.01	0.87
	Number Layers	[2,4]	1	3
	Circuit Copies	[2,3]	1	3
	Number of optimization runs		**95**	

to the previous results, the quantum REINFORCE method is the first to observe the first and the fourth to optimal solutions, while Q-learning is faster for the third and the last two solutions. Therefore, all six Q-learning agents converge to the optimal solution faster than the quantum REINFORCE agents. Nevertheless, the quantum REINFORCE method is the first to obtain an optimal solution. Classical REINFORCE needs more training episodes again to obtain the first solution and needs more than 6,000 additional training episodes to solve the problem compared to the quantum method.

Quantum REINFORCE finds the optimal solution after 1,600 iterations, while Q-learning obtains the first optimal solution after 2,403 training iterations, and REINFORCE is last with 6,068 iterations. All six Q-learning agents reach the optimal policy after 9,733 iterations, followed by quantum REINFORCE with 13,465 iterations, and lastly, classical REINFORCE needs 19,756 training iterations.

The training of Q-learning is significantly faster than the other two methods. However, the quantum method shows a better performance in the initial training phases, where two-thirds of the agents solve the problem faster than Q-learning.

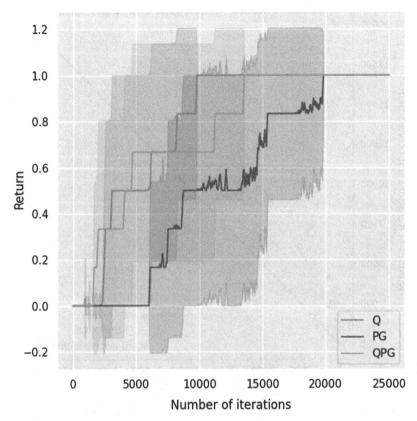

Figure 7.3 Training results for Q-learning, REINFORCE and quantum REINFORCE for the 2D8 problem

However, in total, the quantum REINFORCE algorithm needs significantly longer than the Q-learning algorithm. Comparing the curve for the two REINFORCE methods shows a very similar behavior, with the only difference that the quantum method learns faster with more than 6,000 training iterations difference. Furthermore, the training of the quantum method is more stable since its training curve is smoother than the classical method.

To factor out the randomness until the first successful training episode, which starts the training. A similar analysis to the 2D4 case is performed. In fig. 7.4 the

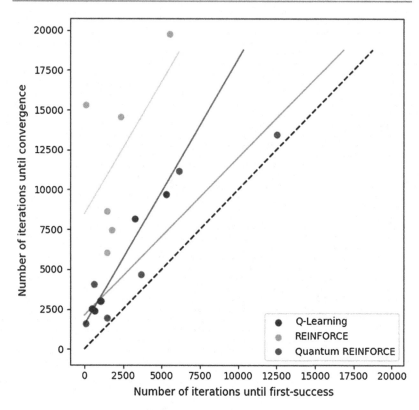

Figure 7.4 Scatter-plot of the iteration of the first-success against the iteration of convergence for the three algorithms on the 2D8 problem instance, with a least-squares regression on the data points for comparison to the theoretical optimal first-success to convergence ratio

iteration with the first observation of the goal state and the iterations until convergence is plotted. Similar to the 2D4 instance, the quantum method shows a close to optimal training behavior since the regression line is close to the optimal convergence behavior. Interestingly, although Q-learning seemed to solve the problem faster than the quantum method in fig. 7.3, removing the initial random behavior in fig. 7.4 reveals that the quantum method clearly shows a better convergence rate. Again the plateauing behavior of the training results in fig. 7.3 of the quantum algorithm is explained by the larger number of iterations until the first successful episode is obtained.

Table 7.4 Comparing mean and standard deviation for each algorithm between the first-time reaching the target until convergence and the mean and standard deviation for the number of observed training samples for the 2D8 problem

Algorithm	First-success to convergence		Policy updates	
	Mean	Std	Mean	Std
Q-learning	2,839.83	1,298.81	–	–
REINFORCE	9,862.33	4,197.25	1,300.5	1,222.9
Quantum REINFORCE	2,070.17	1,620.72	45.0	21.44

Similar to the 2D4 problem, the classical REINFORCE algorithm needs the most iterations until it converges. However, comparing it to fig. 7.2 the least-squares model is relatively seen closer to the optimal convergence ratio, although it is still far away from the other algorithms. This is supported by the statistics over the number of iterations between first-success and convergence and the number of observed training samples shown in table 7.4. The mean number of iterations between first-success and convergence in the 2D4 case has been about an order of magnitude smaller in Q-learning and the quantum algorithm compared to the classical REINFORCE, while it is only about five times smaller in the larger problem instance. The classical REINFORCE algorithm needs, on average 9,862.33 iterations, while the quantum method needs 2,070.17 and Q-learning 2,839.83. This reduction might be due to a faster training of the neural network in the 2D8 setting.

Table 7.4 further verifies that, on average, the quantum REINFORCE algorithm converges the fastest to the optimal solution, compared to Q-learning. As in the previous problem-setting, Q-learning has a lower standard deviation than the policy gradient methods, although the difference is lower than before. This is probably due to the increased complexity of the problem, which leads to many more Q-values in the Q-table and hence more updates of these values. This can lead to an increase in the standard deviation.

Finally, the quantum method only needs a fraction of policy updates compared to its classical counterpart. Furthermore, the number of policy updates only doubled for the quantum algorithm from the 2D4 to the 2D8 instance, while the classical method needs about 20 times more updates. The most probable explanation for this phenomenon is the increased size of the neural network, which is analyzed in the following.

The number of trainable parameters shows considerable differences between the three models. The Q-table of the standard Q-learning method has 256 entries. The neural network with a single hidden layer and 45 units has 270 trainable weights plus 49 bias terms, which gives 319 parameters in total. In contrast to those methods, the quantum REINFORCE algorithm has only 60 parameters in total, divided into 36 weights in the QVC and another 24 parameters for the linear model.

The number of parameters in the quantum REINFORCE method is less than twice the number of parameters as in the 2D4 problem, while the other two methods need about four times more parameters. Like in the 2D4, the results support the findings in [Chen et al., 2020] for variational quantum DQL. The quantum method needs less than a quarter of the parameters compared to its classical counterpart. These results show that the quantum method needs fewer training episodes than the classical method and achieves this with only a fraction of the classical REINFORCE parameters.

7.2.3   High Complexity with the 3D Game

The last complexity level for comparing the three algorithms is the 3D6 IG. It has 216 states in total, thus 3.75 times more states than the 2D8 game. Furthermore, since it is in 3D space, six actions can be taken instead of only four, as in the 2D versions. This problem can still be solved with classical Q-learning. Nevertheless, the number of states would usually justify using a neural network-based algorithm like DQN or a policy gradient method instead of Q-learning. The Q-table has 1,296 entries, which can be represented more efficiently with an advanced approximation method. For the 3D6 game, the same steps as in the previous two experiments are done.

The parameter range for the Q-learning algorithm is the following. The learning rate (α) is in the interval [0.2, 0.4], the discount factor (γ) ranges from 0.8 to 0.9 and epsilon is in the range [0.9, 0.99]. The three intervals are discretized with step-size 0.01. This leads to hyper-parameter space with 2,310 possible combinations and 116 optimization iterations to search 5% of the space.

Similar to the 2D4 and 2D8 results, there are multiple optimal parameter configurations for learning rate and discount factor, while the value for epsilon is unique with 0.9. All combinations for the two parameters are shown in appendix B.1 in the electronic supplementary material. The values of 0.25 for the learning rate and 0.83 for the discount factor as shown in table 7.5 are the values used for the comparison in fig. 7.5.

For the hyper-parameter optimization of the classical REINFORCE algorithm, the learning rate is in the interval [0.008, 0.011], discretized with a step-size of 0.0002. The discount factor values range from 0.92 to 0.99 discretized in steps of

Table 7.5 Parameter ranges for hyper-parameter optimization and optimal values for Q-learning, REINFORCE and quantum REINFORCE in the 3D problem

Method	Parameter	Range	Step-size	Optimum
Q-learning	Learning rate	[0.2 , 0.4]	0.01	0.25*
	Discount factor	[0.8, 0.9]	0.01	0.83*
	Epsilon	[0.9 , 0.99]	0.01	0.9
	Number of optimization runs		**116**	
			*multiple optimal values	
REINFORCE	Learning rate	[0.008 , 0.011]	0.0002	0.0108
	Discount factor	[0.92, 0.99]	0.01	0.96
	Number Layers	[1 , 2]	1	1
	Number Units	[10 , 50]	5	45
	Number of optimization runs		**116**	
quantum REINFORCE	Learning rate	[0.2 , 0.35]	0.01	0.23
	Discount factor	[0.8 , 0.9]	0.01	0.85
	Number Layers	[2,4]	1	3
	Circuit Copies	[2,4]	1	3
	Number of optimization runs		**80**	

length 0.01. The neural network has one or two layers, with the number of units in [10, 50] discretized with increments of five. This induces a parameter space with 2,304 possible parameter combinations in total, which requires 116 optimization steps.

The optimal values for classical REINFORCE obtained by Optuna are a learning rate of 0.0108 and a discount factor of 0.96. The neural network configuration that achieves the best performance with these parameters has one layer with 45 units. The results are as well shown in table 7.5.

The parameter space for hyper-parameter optimization in the case of the quantum REINFORCE method is induced by the following parameter ranges. The learning rate ranges from 0.2 to 0.35, and the discount factor is in the interval [0.8, 0.9]. Both intervals are discretized with a step size of 0.01. The QVC layout has 2, 3, or 4 layers and circuit copies. This results in a circuit with a maximum of 12 qubits. Therefore, the parameter space has a total of 1,584 parameter combinations, which require 80 optimization runs.

The optimization with Optuna resulted in an optimal learning rate of 0.23 and a discount factor of 0.85. The corresponding QVC configuration uses three layers and the same number of copies. Therefore, the QVC needs nine qubits. The whole parameter ranges and optimal values are shown in table 7.5.

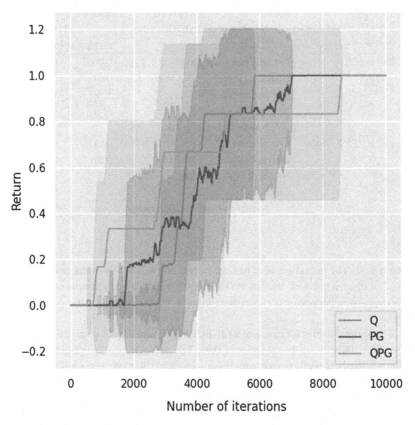

Figure 7.5 Training results for Q-learning, REINFORCE and quantum REINFORCE for the 3D6 problem

Comparing the results of the three methods on the 3D6 IG shows a similar picture compared to the earlier methods. Q-learning converges first, followed by the classical REINFORCE algorithm, and quantum REINFORCE needs the most training. Although all six Q-learning agents solve the problem the fastest, the first four quantum agents need fewer training episodes than Q-learning to solve the IG. Until the fifth agent converged to the optimal solution, quantum REINFORCE and Q-learning show a comparable learning curve. However, the last quantum agent needs significantly more training steps to obtain the optimal solution. Classical REINFORCE is in between Q-learning and the quantum method to obtain the optimal solution. Nevertheless, it shows a steep learning curve for all six agents and eventually converges faster than its quantum counterpart. The learning curves in fig. 7.5 visualize the training results.

The quantum REINFORCE method obtains the first solution after 795 training episodes, followed by classical REINFORCE with 1,741 training steps, and finally Q-learning with 2,839 episodes. The number of training steps until all agents converge is 5,796 for Q-learning, 6,971 for classical REINFORCE, and 8,543 for the quantum method. Thus, although the first quantum agent converged first, it takes 7,748 additional iterations until all agents are trained. This is in contrast to the classical method, which only needs 5,230 episodes between the first and last agents are fully trained.

Although the results from fig. 7.5 indicate that the classical REINFORCE algorithm is superior to its quantum equivalent, the previous sections have shown that removing the early training phase until the first successful episode was observed provides a more accurate picture for the analysis. Therefore, fig. 7.6 shows the number of iterations between the first success and convergence for the three algorithms.

Figure 7.6 reveals a similar picture as before. Q-learning and quantum REINFORCE show a similar learning behavior, which is closer to the optimal convergence rate. Nevertheless, both least-squares predictions are further away from the optimal line than the previous two problem instances. With two exceptions, the quantum algorithm would be close to the optimal learning behavior. However, those outliers blur the picture, and more data points would be required for a better analysis. As in the smaller problem instances, the first-success until convergence analysis shows that the plateaus in the training of the quantum algorithm are due to the late observation of a successful episode that starts the training. In essence, this indicates that in opposition to fig. 7.5 the quantum algorithm shows a superior learning behavior compared to its classical counterpart.

In contrast to the previous two results, classical REINFORCE shows a more stable behavior, with the least-squares prediction being the closest to the theoretical optimum. This is supported by the analysis of the iterations between the first-success and convergence in table 7.6. The quantum algorithm, on average converged the fastest with 2,122.33 iterations, Q-learning needs an additional 600 episodes with 2,729.83 iterations, and REINFORCE needs, on average, one and a half times as many iterations as its quantum versions. This is the least amount of iterations

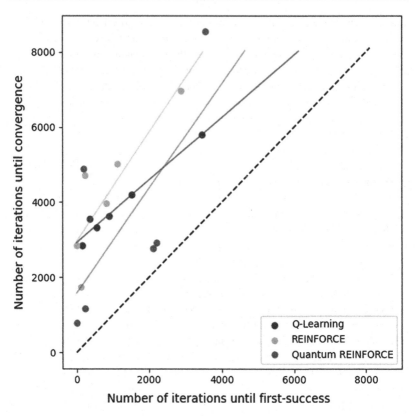

Figure 7.6 Scatter-plot of the iteration of the first-success against the iteration of convergence for the three algorithms on the 3D6 problem instance, with a least-squares regression on the data points for comparison to the theoretical optimal first-success to convergence ratio

between the first successful episode and convergence in all three instances. This is probably due to the increased problem size. It is well known that Q-learning experiences problems with large state spaces from reduced performance due to the tabular value-function representation. Algorithms that use a more sophisticated function approximator like a neural network are usually not as effective in simple problems. They need extensive and complex state spaces to leverage their superior approximation methods.

Table 7.6 Comparing mean and standard deviation for each algorithm between the first time reaching the target until convergence and the mean and standard deviation for the number of observed training samples for the 3D6 problem

Algorithm	First-success to convergence		Policy updates	
	Mean	Std	Mean	Std
Q-learning	2,729.83	247.66	–	–
REINFORCE	3,343.17	946.17	1,093.0	250.07
Quantum REINFORCE	2,122.33	1,920.44	101.67	24.14

The variance within the data is the largest for the quantum algorithm, which again does not fit with the observations from the 2D problem instances. It is about twice as large as its classical counterpart. The number of policy updates until the algorithm converged provides an explanation for this behavior and the outliers in fig. 7.6. The quantum algorithm needs an order of magnitude fewer policy updates and shows again stable values with a low standard deviation.

This indicates that the quantum algorithm still learns fast and is reliable. However, the second source of randomness, which is the random action selection during the training, leads to long times until a successful training sample has been obtained. Thus, the quantum method still needs only a fraction of the training samples to learn the optimal behavior. However, the long paths to the final position in the large problem setting lead to many unsuccessful episodes from which no information can be obtained. Figure 7.6 can not filter out this source of randomness and therefore indicates decreased performance at first sight.

Comparing the number of parameters of each method shows a similar picture as in the previous two experiments. Q-learning's Q-table has 1,296 entries, which corresponds to the number of trainable parameters. The neural network with one layer and 45 units has in total 456 parameters that split up into 51 bias values and 405 weights. The QVC has 54 parameterized circuits and an additional 54 weights in the linear model. With 108 trainable parameters, it needs less than a quarter of the parameters compared to the other two methods. This is in line with the previous observations.

7.3   Discussion

Comparing the results from the three experiments on the different instances of the IG revealed that the quantum REINFORCE method trumps its classical counterpart in terms of the number of training episodes that have to be observed until the training converged. The quantum method needed fewer training episodes in all three experiments until the first of the six agents converged. Furthermore, the quantum algorithm required fewer episodes until all agents converged in all IG instances except the 3D6 IG. In the 3D problem, the last quantum agent needed 819 training steps to converge. This is about 10% of the algorithm's total training time.

In contrast, the classical algorithm required 6,291 more steps than the quantum method. This is about 45% of the algorithms observed training episodes. Despite the increased time in the 3D problem, the quantum algorithm showed overall a better performance. The algorithm was further able to compete with Q-learning, which is a competitive method due to its straightforward training method. The quantum algorithm needed fewer training episodes to observe the first solution in the 2D8 and 3D6 environments. In the 2D4 setting, the quantum method needed two additional episodes compared to Q-learning. However, the quantum algorithm showed, on average, an increased number of observed episodes of about 185% compared to Q-learning. A summary of the minimum and the maximum number of training episodes is shown in table 7.7 for all three methods.

Table 7.7 Minimum and maximum number of training iterations until convergence for each RL algorithm in the three environments

Environment	Iterations	Q-learning	REINFORCE	quantum REINFORCE
2D4	Minimum	**73**	504	75
	Maximum	**310**	929	838
2D8	Minimum	2,403	6,068	**1,600**
	Maximum	**9,733**	19,756	13,465
3D6	Minimum	2,839	1,741	**795**
	Maximum	**5,796**	6,971	8,543

Although the quantum method needs more training episodes than the Q-learning, the results still show that it is superior compared to its classical counterpart in terms of convergence speed. Therefore, it can safely be said that QVC can be used in policy gradient RL. Furthermore, in the quantum REINFORCE algorithm, the QVC increases the algorithm's performance compared to its classical equivalent with a neural network.

The analysis further showed that only considering the training results provides a skewed picture since it does not take the randomness from the initial training phase into consideration. Therefore, rescaling the data to take the time between

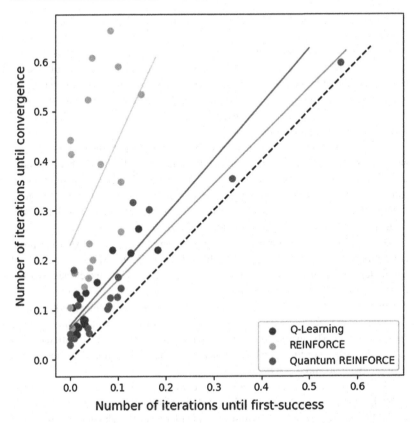

Figure 7.7 Combined scatter-plot of the iteration of the first-success against the iteration of convergence for the three algorithms on all three problem instance, with a least-squares regression on the data points for comparison to the theoretical optimal first-success to convergence ratio

the first successful episode and the algorithm's convergence to the optimal solution provides a better picture for analyzing the convergence rate. The quantum algorithm has shown superior convergence performance for all three instances compared to the other two algorithms. Only in the 3D6 problem setting suffered the quantum algorithm from higher variance in the average number of iterations between first-success and convergence. Figure 7.7 summarizes the results by combining the data from all three settings into one picture by normalizing the number of iterations. It further generalizes the results to provide a prediction independent of the specific problem setting.

The figure clearly shows that the quantum algorithm generally performs the best. It converges the fastest of all the three algorithms and only has a few outliers. As already discussed in the previous section, the outliers are due to the second source of randomness in training RL. This is shown by the fact that the quantum algorithm needs fewer policy updates compared to its classical counterpart and has a lower variance within the number of updates.

The number of trainable parameters as the second comparison criterion showed an even more significant advantage of the quantum method. The number of parameters directly corresponds to memory usage. It will show if quantum methods in RL can reduce the required memory usage. In the small problem setup (2D4) with 16 states and four actions, the quantum needed 32 trainable parameters, which is 50% of the parameters used by Q-learning and 43% of the classical method.

The trend of fewer parameters becomes evident in the more extensive problem settings. In the 2D8 environment with 64 states and four actions, the number of parameters increased to only 60, while Q-learning required 256 and classical REIN-FORCE 319. This means that the quantum method needs only 23% of Q-learning's parameters and 18% of its classical counterpart, a significant memory reduction. In the complex 3D scenario with 216 states and six actions, the quantum method used 108 parameters. The increased number of parameters is due to the additional qubit to encode the Z-coordinate on the 3D map since the QVC layout does not change between the 3D6 and the 2D8 problem.

The 108 parameters are only 8% of Q-learning's and 23% of classical REIN-FORCE's memory consumption. Thus, with increased problem size, the quantum method's memory growth is significantly smaller than the classical method. The results are summarized in table 7.8.

The above shows that the quantum method is memory efficient. Combining this with the improved convergence, the first research question is more than posi-

Table 7.8 Number of trainable parameters for each RL algorithm in the three environments

Environment	States	Actions	Q-learning	REINFORCE	quantum REINFORCE
2D4	16	4	64	74	32
2D8	64	4	256	319	60
3D6	216	6	1,296	456	108

tively answered. The QVC was successfully implemented in the Monte-Carlo REIN-FORCE method and improved convergence speed and significantly reduced memory usage. These results make it likely that similar results can be obtained for other policy gradient methods. Furthermore, the reduced memory shows the possibility that quantum methods need less computational power to achieve the same or even better results than classical methods. This result is in line with the reduced memory usage in DQN with a QVC obtained in [Chen et al., 2020]. In a different setting, [Beer et al., 2020] have shown that quantum approximators are memory efficient. However, they did not compare the memory usage to a classical neural network achieving the same results as the above experiments.

The results from this chapter should answer the first research question stated in section 4.4, which is the following:

Can QVC be utilized to approximate the policy in policy gradient RL methods directly, and how do they perform compared to their classic counterparts?

The previous section's results show that the quantum REINFORCE algorithm, which utilizes a QVC to approximate the policy, can solve both 2D and the complex 3D6 IG problem, which answers the central part of the research question. For the second part, which concerns the performance of the novel quantum algorithm, the convergence rate and the memory consumption of the function approximator have been analyzed. This analysis revealed that the quantum algorithm developed in this work has a superior convergence rate compared to its classical counterpart and a slightly improved convergence behavior compared to the classical value-based Q-learning algorithm. These findings indicate that utilizing QVC in policy gradient RL boosts the convergence speed of the algorithm.

Memory consumption as the second performance criterion shows similar results. In the classical setting, the function approximator's size grows linear with the state space. The quantum algorithm shows, at first sight, an unintuitive but very promising behavior. With the problem's complexity increasing, the QVC needs relatively fewer parameters, making it a memory-efficient learning method. Therefore, the quantum REINFORCE algorithm seems to be a good candidate to solve the problem of requiring massive computation resources to tackle complex RL problems. Once large enough quantum hardware can be used, the novel algorithm can be applied to solve extremely complex problems, which would classically require enormous computational power. The next chapter will analyze the second research question to identify the possible benefits of learning on real quantum hardware.

Evaluating Quantum REINFORCE on IBM's Quantum Hardware

8

This chapter aims to answer the second research question if the quantum REIN-FORCE method can be trained on today's NISQ devices without a quantum simulator. If this is possible, it raises the question, if full quantum learning provides benefits, like improved convergence, compared to learning with a simulator.

These questions are going to be answered in two steps. In a first approach, the agents are trained on a simulator until the optimal solution is obtained. The trained QVCs parameter values are then evaluated on a real quantum computer to check if they still solve the problem. This first step will reveal if today's erroneous quantum hardware can reproduce the same results as a perfect quantum simulator.

In the second step, the whole training is performed on the quantum computer. This experiment will demonstrate if the quantum REINFORCE RL algorithm will yield the same results on a real device instead of the simulator. It will further reveal obstacles that need to be taken into account in the future when quantum computers are broadly available for research and application.

8.1 Evaluating the Trained Algorithm

In the first step, the quantum REINFORCE agents are trained with a simulator, similar to the last chapter. The obtained parameters for the QVC are then transferred to a quantum computer to evaluate whether the noisy quantum hardware can obtain the same results. For the evaluation, each agent repeats the 2D4 game five times. The training is performed with the optimal hyper-parameters from the previous chapter. Thus, the QVC has two trainable layers and is copied twice, which leads to a four-qubit quantum circuit. The small number of qubits allows choosing from various of IBM's quantum computers since most have at least five qubits.

© The Author(s), under exclusive license to Springer Fachmedien Wiesbaden
GmbH, part of Springer Nature 2022
L. Kunczik, *Reinforcement Learning with Hybrid Quantum Approximation
in the NISQ Context*, https://doi.org/10.1007/978-3-658-37616-1_8

The experiment was performed on the *ibmq_rome* system. It is a five-qubit quantum processor based on the Falcon chip family. The Falcon chip-set was introduced in 2019 and is frequently improved next to the development of new chips with more qubits, and reduced error rates [IBM, 2021a]. The ibmq_rome system used in this study is powered by a Falcon process from the 4th revision. IBM improves the systems gradually over time. Therefore, the system-specific information is provided for comparability with other systems in the future.

Quantum Volume is a term introduce by IBM [Cross et al., 2019] as a performance measure for quantum computers. It connects the chip size (number of qubits) with its error rate to have a single number that provides information about the computational capabilities of a quantum computer. The ibmq_rome quantum system has a quantum volume of $32 = 2^5$, which means that a quantum circuit with at most five gates on each qubit will yield the exact results.

The last hardware-specific part to mention is the topology of the chip. It defines the connections between the qubits and therefore determines which qubits can interact with each other. The ibmq_rome chip has a line topology as shown in fig. 8.1. This means that each qubit is only connected to its direct neighbor.

Figure 8.1 Topology of the ibmq_rome quantum computer

The topology directly influences the quality of the results. If the algorithm uses CNOT gates between each qubits, the line topology requires that the qubits be adjacent to perform the CNOT operation. This usually results in longer circuits that need to swap the qubits' position on the chip such that they are directly connected. For example, for a CNOT gate between the first and the third qubit of the ibmq_rome chip, one qubits need to change places with the second qubit to be directly connected to the other qubit. The swapping operations prolong the circuit, which leads to a higher probability of errors in the result [Rieffel and Polak, 2011].

The advantage of the QVC for the 2D4 IG is that only two of the four qubits need to be directly connected, satisfied in the line topology. Thus, no swap operations are needed, which increases the quality of the result. One example circuit used during the evaluation is given in fig. 8.2. It shows only pairwise interaction between the qubits and that the fifth qubit is not used. The blocks, e.g., encoding and trainable layers of the QVC, are not easily recognizable. This is because a compiler optimizes

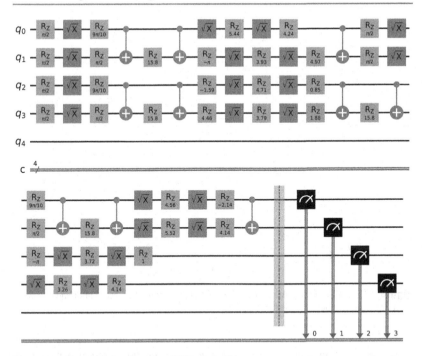

Figure 8.2 Example circuit on the QVC on the ibmq_rome quantum computer taken from the IBM Quantum Experience. The circuit was split into two parts for better visualization

the algorithm sent to the quantum hardware to reduce the number of used gates and increase the algorithm's robustness.

The QVC was computed with 2,000 shots for the evaluation, meaning that the quantum circuit is computed 2,000 times. The results are averaged to obtain a good statistic over the state distribution. This number provides satisfactory results while reducing the computational time (a full analysis of this is given in table 8.2). The evaluations results are shown in fig. 8.3. The QVC is robust enough such that all six agents still solve the problem when the computation is performed on an erroneous quantum computer instead of a perfect quantum simulator. This shows that the quantum REINFORCE algorithm can deal with the quantum hardware errors without any deterioration in the performance once the optimal solution was obtained.

However, if the algorithm evaluated on the quantum computer solves the problem in the first of the five evaluation runs, the result will not change for the other evaluations. This is due to the algorithm's implementation since it stores the results from the QVC after a new state was visited, as described in algorithm 6.

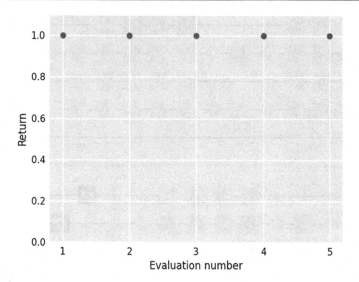

Figure 8.3 Results of six quantum REINFORCE agents trained on a simulator and evaluated on a quantum computer, with stored policy

The policy storage should be cleared after each evaluation iteration to obtain reliable results. Therefore, the same computations are repeated, and the policy storage is emptied after each evaluation run. This change reveals if the previous results are due to chance since only a single computation is performed if a state in the 2D4 IG is visited for the first time. Without the policy storage, the QVC has to be computed in each of the five evaluation runs, and it will reveal fluctuations in the results.

The evaluation results visualized in fig. 8.4 show that the QVC is, in fact, robust enough to achieve the optimal solution consistently. All six agents take the optimal path to the goal in each of the five evaluation runs and have thus used an optimal policy. The results show that the quantum REINFORCE method with a QVC can handle quantum hardware with errors and is still robust enough, to achieve the same evaluation results as the perfect simulator.

This is similar to the findings in [Chen et al., 2020]. They performed a similar experiment with the variational quantum DQL approach. However, the variational quantum DQL algorithm reached the optimal solution only in four out of seven tests. The other three obtained a sub-optimal solution with one additional step. The slightly better results in this study might be due to improved quantum hardware since they performed their evaluations.

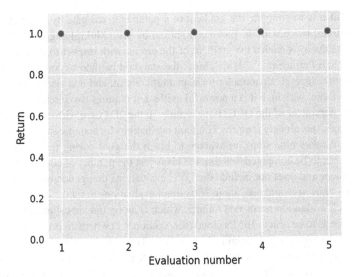

Figure 8.4 Results of six quantum REINFORCE agents trained on a simulator and evaluated on a quantum computer, without stored policy

8.2 The Full Training Loop on the Quantum Hardware

The above experiments have shown that once the QVC is trained, it will yield the same result on a quantum computer. The next question is whether it is possible to perform the full training on the quantum hardware instead of the exact simulator and still obtain the same results.

In order to transition the training onto the quantum computer, the algorithm needs to be optimized. One of the main issues that needs to be tackled is computing the gradient during the training. On a classical computer, automatic differentiation frameworks compute the gradients of a given function [Abadi et al., 2016]. This is one of the features that PennyLane [Bergholm et al., 2018] provides if a simulator is used. However, once the computations are transitioned to a quantum computer, the parameter shift rule is used. The parameter shift rule is a method to compute the gradient of a circuit on a quantum computer.

The idea is to compute the gradient of a parameterized gate by evaluating the whole circuit twice with the parameter shifted by $\pm\frac{\pi}{2}$. Multiplying the difference of the results by $\frac{1}{2}$ yields the gradient of the circuit with respect to the parameter [Schuld and Petruccione, 2018]. This is the standard method to compute the gradient with respect to a parameter for a quantum circuit, and it is also implemented in PennyLane. Although it is a powerful method, it requires two circuit evaluations for each parameter of the QVC. For example, in the 2D4 case with a QVC with 16 parameters, this already requires 32 circuit evaluations for a single sample. Assuming that it takes nine steps on average to reach the goal during the training, the QVC needs to be computed 288 times. This is just the number of computations for the gradient and does not include the QVC evaluations that generated the training episode. Thus, it would take about 300 circuit evaluations in total.

If each evaluation needs one minute, which is an optimistic estimate, since the circuit needs to be sent to IBM's cloud (see. section 9.1) where preprocessing steps, like the compilation and optimization, is done. Thus, one training iteration would need about five hours. Now taking into account that multiple training episodes are needed, some number between 75 and 838 iterations for the 2D4 Information Game (fig. 7.1), the training could take up to half a year in the worst case and about two weeks in the best case. These numbers are only for one agent. If the training would be performed with six agents like in the previous chapter, the whole process becomes intractable.

This small example highlights the major problem if the training should be performed on a real quantum computer. It further shows the need to reduce the number of circuit evaluations in order to make the training process applicable to quantum hardware.

One approach to reducing the number of circuit evaluations on the quantum computer is to perform gradient computations with a simulator. This is more efficient than the parameter shift rule since automatic differentiation can be used, which only requires computing the QVC once. This leads to an algorithm that performs the policy evaluation to generate an episode on the quantum computer. The costly derivatives to update the QVC are done with a simulator. The algorithm that performs these steps is given in algorithm 7. It is to note that this adaptation could be introduced as well for any other quantum RL algorithm to achieve a training that in-cooperates real quantum hardware.

Algorithm 7 The quantum REINFROCE algorithm with a QVC for attacker-defenderr scenarios and stored policy for training on a quantum computer.

Require: θ: A weight vector for the QVC with the values initialized to one
Require: $\pi_{QC}(\cdot|\cdot)$: A QVC-policy computed on a quantum computer
Require: $\pi_S(\cdot|\cdot)$: A QVC-policy computed on a quantum simulator
Require: $\tilde{\pi}$: An empty policy-storage (e.g. dictionary)
Require: $\gamma \in (0, 1]$: Discounting factor
Require: $N \in \mathbb{N}$: Number of iterations
  for $n = 1, \ldots, N$ **do**
    Generate an episode $\{S_0, A_0, R_1, \ldots, S_{T-1}, A_{T-1}, R_T\}$ from $\tilde{\pi}$; if $\tilde{\pi}$ does not hold an
    entry for the given state, query π_{QC} and store the result in $\tilde{\pi}$
    if $\sum_{k=1}^{T} R_k \geq 0$ **then**
      for $t = 0, \ldots, T - 1$ **do**
        $G \leftarrow \sum_{k=t}^{T} \gamma^{k-t-1} R_k$
        $\theta \leftarrow \theta + \alpha \gamma^t G \nabla \ln \pi_S(A_t|S_t, \theta)$
        Empty $\tilde{\pi}$
      end for
    end if
  end for

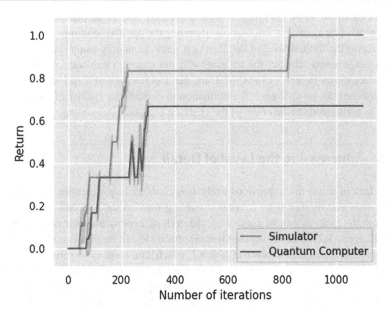

Figure 8.5 Training results for the quantum REINFORCE agent on the 2D4 problem on the ibmq_rome quantum computer

The above algorithm does not realize full quantum training. However, it is a trade-off that still performs the critical step that derives the policy on the quantum computer. The whole training with the optimal parameter values for the 2D4 IG from table 7.1 and six agents is repeated with algorithm 7, to test whether the training on the noisy quantum hardware can compete with the results obtained from a simulator. The training is performed on the ibmq_rome quantum computer with the number of evaluation shots set to 2,000. The obtained results are shown in fig. 8.5. The results from the simulator obtained during the experiments in the last chapter are included for better comparison.

The results reveal that four out of six agents obtained the optimal solution. The other two agents have not been able to solve the problem in 1,100 training iterations. By comparing the results with the previous results, the two algorithms show a similar learning behavior. The learning curves for the first two agents are almost identical. The only difference is that the training on the quantum computer needs more iterations than the simulator agents. Until episode 250, the learning curve from the quantum computer is almost comparable to the simulator results. However, the results start to deviate afterward. The results start to fluctuate until the next two agents solve the problem around iteration 300. From there onwards, the two remaining agents are not capable of finding the optimal solution.

Due to the limited level of detail, which stems from only using six agents, the question remains whether the training with the quantum hardware results in an accuracy closer to 50% or to 90%? Thus, the data from a single run with only six agents does not provide enough information to evaluate the quality of the learned results nor to analyze the effect of the quantum computer.

8.3   Increasing the Level of Detail

The lack of detail in the previous analysis is countered by increasing the number of agents and the number of repeated training runs on the real quantum device. Before the analysis can be conducted, the optimal hyper-parameter configuration for 12 agents is obtained using the Optuna software like in the previous chapter. The hyper-parameter space is shown in table 8.1, which is the same as in section 7.2.1. For a detailed discussion on hyper-parameter space and hyper-parameter optimization, the reader is referred to the previous chapter whit special attention on section 7.2.1.

The optimal hyper-parameter configuration for the 12 agents in the 2D4 problem as shown in table 8.1 differs slightly from the configuration for six agents. Both learning rate and discount factor increase. The new learning rate for the 12 agent setting rose from 0.43 to 0.5, while the discount factor grew from 0.95 to 0.96

compared to the six agent case. The increase in learning rate and discount factor leads to large value updates in the QVC in each training step. Therefore, the training should converge faster.

The most significant finding from repeating the optimization is that the layout of the QVC does not change. The 12 agent setting uses as well two layers with two circuits. This indicates that the QVCs' layout is the optimal configuration that represents best the policy in the 2D4 IG.

The differences in the optimal hyper-parameter setting are due to the increased number of agents since the purpose of the optimization is to fit the algorithm to the particular problem instance. By changing the number of agents, a different set of random numbers will be used for the new agents, which lead to slightly altered values. However, the layout of the QVC does not change, and 32 parameters seem to be the minimum number of parameters that can solve the problem reliably.

To compare the 12 agents to the six agent setting, fig. 8.6 shows the training curves of both configurations. As expected, the two graphs show similar behavior, with the 12 agents' curve being smoother due to the increased number of agents. The slightly reduced iterations until all agents converged for the 12 agent setting are due to the accelerated learning speed from the increased learning and discounting factors. Again, this is not significant, since overall, the curve lags behind the six agent configuration, indicating a slower training, as with the reduced number of agents.

Before performing the training on the quantum hardware, the trained QVC is evaluated on the quantum computer to verify that the results match the expected values. Unfortunately, the ibm_rome quantum computer was retired before all experiments could be performed.

Table 8.1 Parameter ranges for hyper-parameter optimization and optimal values for the quantum REINFORCE algorithm in the 2D4 problem with 12 agents

Method	Parameter	Range	Step-size	Optimum
Quantum REINFORCE	Learning rate	[0.35 , 0.55]	0.01	0.5
	Discount factor	[0.85 , 0.99]	0.01	0.96
	Number Layers	[2,4]	1	2
	Circuit Copies	[1,2]	1	2
	Number of optimization runs			**95**

Therefore, some computations are performed on the ibm_bogota system. It is also a five-qubit quantum computer with the same layout, chip family, quantum volume and similar error rates as the ibm_rome system. This should ensure that the results are comparable since they have been performed on systems with almost identical characteristics.

The trained algorithm is evaluated ten times, while the policy storage is cleared after each evaluation. The results of this experiment are shown in fig. 8.7. These results are as expected from section 8.1. All agents perform optimally when the trained QVC is evaluated on the real quantum hardware. This only partially answers the question if the evaluation of the QVC is stable on the quantum hardware since fig. 8.7 only shows that the overall result does not diverge from the optimal solution. Therefore, the error between the optimal policy on the simulator and the quantum hardware for the optimal path should be analyzed next.

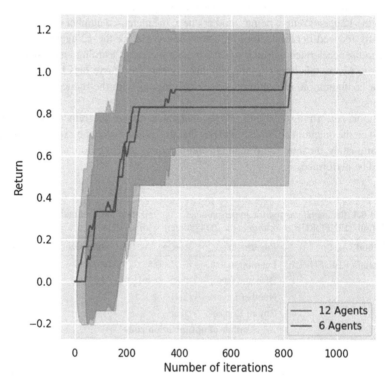

Figure 8.6 Training results for the quantum REINFORCE algorithm with 12 agents on the 2D4 problem

To measure the error between the optimal policy from the simulator and the quantum computer for the optimal path, the mean squared (MSE) between the simulators' policy and the policy obtained from the ten evaluations from above are computed. The results are shown in table 8.2. The table further includes comparing the evaluated number of circuits to compute the QVC and on which quantum computer the computations have been performed. It thus shows how many times the circuit needs to be measured to provide a reliable statistic about the quantum state. The optimal path with 2.000 circuit evaluations is performed on both devices to check whether the devices provide similar results or not.

This analysis is performed to obtain the number of circuit evaluations needed for the experiments. There are no known formulas or rules to determine the optimal number of evaluations. IBM uses by default 1024 evaluations which is an arbitrarily chosen value [Abraham et al., 2019]. There exists some research on determining the number of circuit evaluations for gradient computations in variational algorithms [Sung et al., 2020, Sweke et al., 2020]. However, the results are not comparable to the setting considered in this work since the gradient computations do not need to be as accurate as the actual computation. Therefore, the above analysis was performed to obtain insight into the dependence between accuracy and circuit evaluations.

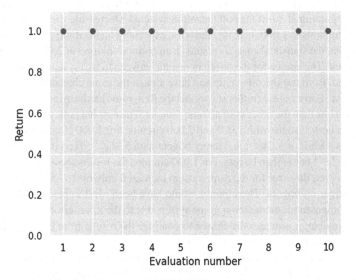

Figure 8.7 Results of the 12 quantum REINFORCE agents trained on a simulator and evaluated on a quantum computer, without stored policy

Table 8.2 Comparing the mean squared error of the optimal policy on the optimal path between the values obtained from the simulator and the results from the quantum hardware evaluated with different numbers of circuit repetitions

Device	ibmq_rome				ibmq_bogota	
# Circuits	8,000	4,000	2,000	1,000	2,000	1,000
MSE	$3.65e^{-4}$	$3.68e^{-4}$	$3.84e^{-4}$	$3.92e^{-4}$	$5.86e^{-4}$	$6.38e^{-4}$

The results show that with decreasing number of circuit evaluations, the MSE increased. Thus, the results are more accurate with more circuit evaluations which make sense in terms of the law of large numbers. The law of large numbers states that by repeating an experiment more often, the results should be closer to the actual expected values. Therefore, by including more samples when computing the quantum state, the closer the state gets to the actual state. The results in table 8.2 further show that 2,000 circuit evaluations are enough to still obtain accurate results whit a reduced number of circuit evaluations. The difference in the MSE is less than $2e^{-5}$, which is small enough to not impact the results. The results with 4,000 executions are even closer to the reference value with the maximum number of evaluations (8,000). However, 2,000 evaluations are an acceptable trade-off between accuracy and computational effort for both quantum systems. Decreasing the number of circuit evaluations to 1,000 increases the error, especially on the ibmq_bogota system. Therefore, the statistical analysis results in an optimal solution of 2,000 evaluations per circuit. The source for the different results between to two systems could not be identified. Both use the same chip and have almost the same characteristics.

The MSE in table 8.2 further shows that the ibm_rome hardware achieves slightly better results than the other system ibmq_bogota. The optimal policy has been evaluated on both systems with 1,000 and 2,000 circuits. For 2,000 circuits ibmq_rome achieves a MSE of $3.84e^{-4}$ and ibmq_bogota with $5.86e^{-4}$. Thus the rome system is about $2e^{-4}$ better than bogota. For 1,000 circuits, the results show similar behavior. However, the error for the rome system increased only by $1e-5$ while it is five times larger on bogota. Therefore, it would have been ideal to perform all subsequent computations on the ibmq_rome system due to the lower error. Although this is not possible, it provides the chance to compare the results from both systems to get more insight into increased error rates.

The evaluated policy on the simulator and the two quantum devices is shown in fig. 8.8, 8.9 and 8.10 for each agent to gain more insight into the errors. This allows a direct comparison between the observed results for each device, and large deviations

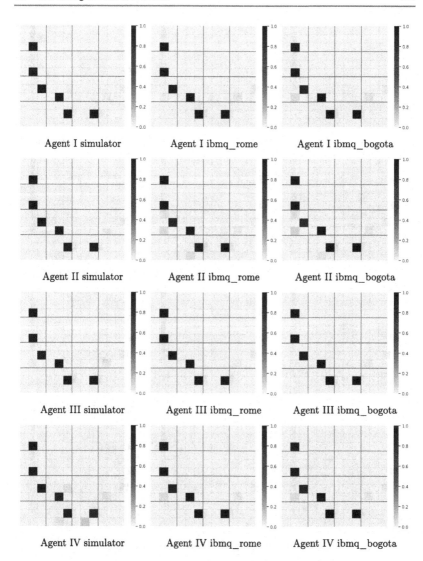

Figure 8.8 The learned policy evaluated on the simulator, ibmq_rome and ibmq_bogota for the optimal path for agent I–IV

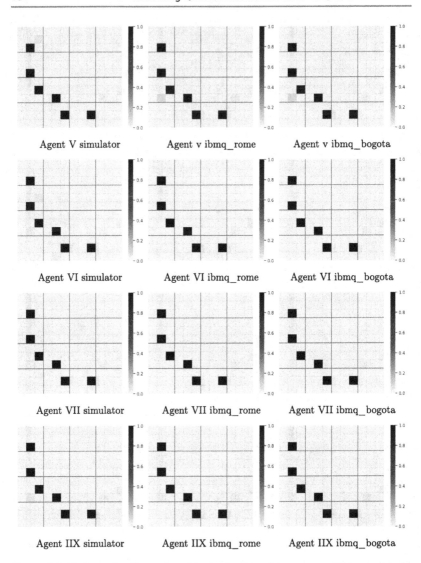

Figure 8.9 The learned policy evaluated on the simulator, ibmq_rome and ibmq_bogota for the optimal path for agent V–IIX

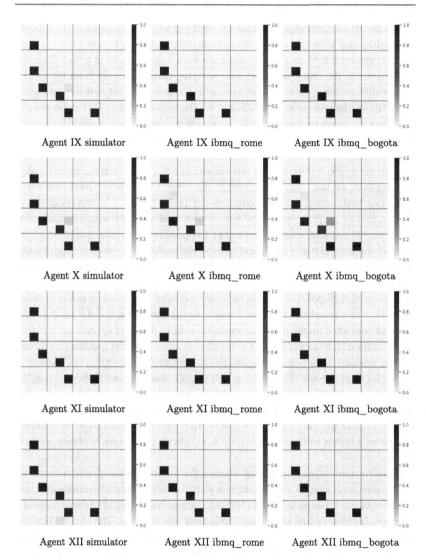

Figure 8.10 The learned policy evaluated on the simulator, ibmq_rome and ibmq_bogota for the optimal path for agent IX–XII

can be pointed out directly. The results from the quantum devices are averaged over ten evaluations. With this representation, the observed deviations in MSE between the two quantum devices for the evaluation with 2,000 circuits in table 8.2 can be explained. On the ibmq_bogota device deviates agent X largely from the results obtained from ibmq_rome and the simulator. Excluding Agent X from computing the MSE results in a value of $4.8257e^{-4}$ for the bogota device. This value is $1e^{-4}$ less than the previous results, while at the same time increase the MSE by $1e^{-5}$ for ibmq_rome. Thus, the large deviations are primarily due to one faulty value in Agent X on the bogota computer.

Comparing the results from the three devices further shows that the approximated policies from the two quantum computers are closer to each other than to the simulator. The MSE between the two quantum devices is $2.8152e^{-4}$. This result shows that the two quantum computers provide almost similar results, which indicates that better results could be obtained if the QVC is optimized to the real quantum system instead of the simulator. This should be explored in the future when the entire training and the hyperparameter optimization can be performed on the quantum system. The above analysis only concerned the MSE of the three policies. However, it does not ensure that all policies will lead to correct results. Therefore, it is left to check that all policies can obtain the correct result. As depicted in fig. 8.8, 8.9 and 8.10, the policies computed on the quantum hardware lead to the same optimal policy, with some minor differences compared to the one obtained from the quantum hardware.

The artifacts do not change the overall behavior since the argmax is taken over all values to determine the chosen action in each state. It further showed that the difference between the two quantum systems is smaller than the simulator, indicating that better results could be obtained if the hyper-parameters are better tuned for the quantum hardware. The results show that for all further computations on the quantum computer, it is thus sufficient to use only 2,000 circuit evaluations to obtain accurate results.

Now the main open question of whether the training on the quantum hardware leads to results that are closer to an accuracy of 50% or 90% from the previous section can be answered. The quantum REINFORCE algorithm is trained 12 times on IBM's quantum hardware using algorithm 7 to answer this question. The obtained results can be used to evaluate if the computational errors will lead to larger fluctuations in the training process to get a clear picture of the algorithm's accuracy on the quantum computer. The training result of each agent is shown in fig. 8.11, 8.12 and 8.13. Training runs one to six have been trained on the ibmq_rome device, and the remaining training runs on the ibmq_bogota system.

The differences between the quantum hardware and the simulator raise the question of any patterns within the training results that yield information about the source

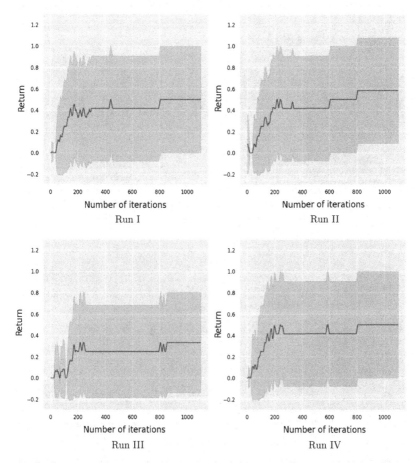

Figure 8.11 Results of the first four runs of the quantum REINFORCE with 12 agents trained with a simulator and evaluated on the ibmq_rome quantum computer, with stored policy

of the deviations. The MSE between the simulator's results and the average results of the 12 training runs for each agent is computed to analyze the differences. The results are shown in table 8.3.

Table 8.3 further provides the MSE for the training runs performed on each device to analyze for differences between the devices. Those values are not as reliable as the combined value since it uses only six of the 12 runs, but it still can provide some insight on the problems during the training.

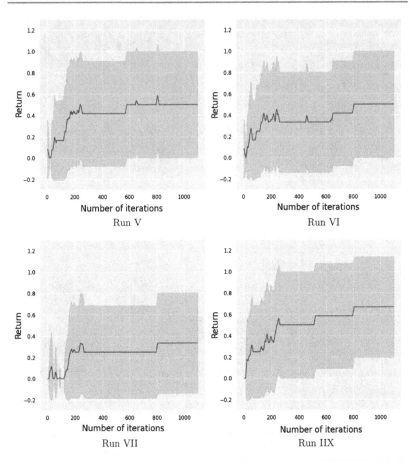

Figure 8.12 Results of the fourths to eights runs of the quantum REINFORCE with 12 agents trained with a simulator and evaluated on the ibmq_rome (Runs V and XI) and the ibmq_bogota (Runs VII and IIX) quantum computer, with stored policy

The combined value show that agents: I, III, IV, V, and X diverge the least from the simulator's values. This means that those agents learn the optimal behavior in most of the 12 training runs. However, agents V and X are the best and obtain in ten out of 12 runs the optimal solution. The other three are only in 75% of the runs successful. On the opposite side is Agent II, with the most significant error and no success in any of the 12 runs. The other agents have at least four training runs in which they obtain the optimal solution.

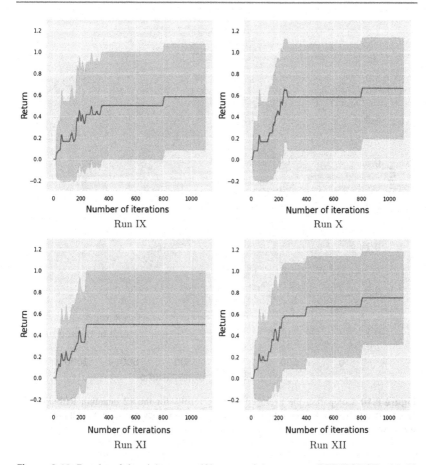

Figure 8.13 Results of the eights to twelfths runs of the quantum REINFORCE with 12 agents trained with a simulator and evaluated on the ibmq_bogota quantum computer, with stored policy

If this analysis is split up between the two devices, the ibmq_bogota device performs better than ibmq_rome. Additional analysis for the differences could not reveal an explanation why ibmq_bogota performs now better compared to the rome device. This can already be seen from the MSE computed over all agents. However, if analyzing the agents separately for the rome quantum computer agents: I, IV, V, VI, and X perform best. Agent X learns the optimal solution in each training run while agents V and VI only in five out of six and I and IV only in four out of six runs.

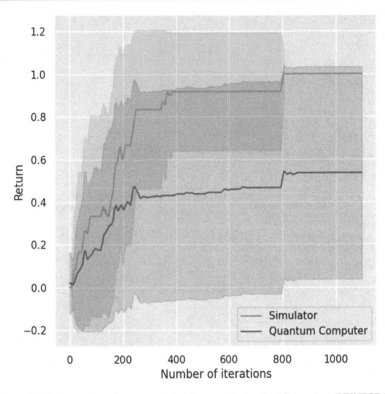

Figure 8.14 Comparing the averaged training results for the 12 quantum REINFORCE agents on the 2D4 problem on the ibmq_rome and ibmq_bogota quantum computer against the simulator results

For the ibmq_bogota computer, the results are a bit different, with agent III learning optimally, agents: I, IV, and V being successful in five out of six runs, and agents: XII, IX, X, and XII learning the solution in 4 training runs. Fig. 8.14 compares the training results for the 12 agents trained on the quantum computer against the simulator. The results show that the quantum hardware achieves an accuracy of 55%.

The above analysis could not identify the exact reason for the sub-optimal solution. However, it shows that there are some differences between the used quantum computers. Contrary to the results in table 8.2 where the ibmq_rome quantum computer showed smaller deviations from the optimal policy, the training results show a completely different picture, which means that no systematical larger error with

Table 8.3 Comparing the mean squared error between the averaged values obtained from the quantum hardware and simulator for each agent, with the combined results for both devices and each separately

Agent	MSE Combined	MSE ibmq_rome	MSE ibmq_bogota
1	0.0164	0.0289	0.0077
2	0.8446	0.84508	0.8463
3	0.0796	0.2191	0.02306
4	0.0898	0.1465	0.0526
5	0.0379	0.0415	0.0357
6	0.3694	0.0702	0.9364
7	0.2181	0.3942	0.0998
8	0.4618	0.7884	0.2257
9	0.1832	0.2460	0.1338
10	0.0772	0.0384	0.1495
11	0.2877	0.2727	0.3155
12	0.3428	0.7739	0.0854
Total	0.2507	0.3221	0.2426

one of the devices could be identified. The most plausible explanation for this phenomenon is that the computational error of the devices varies over time. The quantum computers are daily calibrated to remove errors from environmental changes [IBM, 2021c]. Therefore, the systems' accuracy varies over time, which can lead to the observed results.

Finally, there is still the open question of why the training performed suboptimal. There are two perspectives from which this question can be approached. The first direction is the hardware side, meaning that today's quantum hardware cannot perform reliable computations. Thus, it has too large errors to provide accurate results. The above computations do not utilize error mitigation methods which could improve the results [Roffe, 2019]. The other direction to approach the issue is from the algorithmic side. As already discussed above, it is not practical at the moment to perform all computational steps, e.g., derivatives, on the quantum hardware. This is likely to provide some errors during the weight update, which results in sub-optimal solutions. The other source is the chosen hyper-parameters. It is possible that a different hyper-parameter combination could improve the training on the quantum hardware. For example, a lower learning rate could reduce the error propagation from the derivatives. Thus, many things could alter the solution, but today's quantum computers do not provide the capabilities to explore them in detail.

8.4   Summary and Answering the Research Question

The previous chapter has shown that the quantum REINFORCE method is beneficial compared to its classical counterpart. It needs less trainable weights in the QVC than the classical neural network, which is in line with [Chen et al., 2020], and it converges faster. Those experiments had been performed on a quantum simulator since the access to real hardware is limited, and only a negligible number of computations can be done per day. As argued in the chapter's beginning, the entire algorithm can not be performed on the quantum computer due to this limitation, primarily due to the costly gradient computation. Therefore, algorithm 7 uses a simulator to compute the gradients and performs the evaluation of the QVC on the real device. Although this does not provide the complete picture of the quantum algorithm's performance, it is the first step to get some insights about it, and most importantly, it can already be done on today's hardware.

Before the algorithm was trained on the quantum computer, the pre-trained algorithm was evaluated on the hardware in a first step to evaluate if today's NISQ devices can reproduce results obtained with a simulator. This analysis was split into two steps. In the first step, the agent used the policy storage during evaluations, and in the second, the policy storage was cleared after each episode. In both cases, all agents performed optimally, which showed that evaluating a trained quantum REINFORCE agent on real quantum hardware with noise does not introduce errors that lead to different behavior.

Performing the training on the quantum computer with the same hyper-parameter setting as in the previous chapter showed that not all agents converged to the optimal solution. Four out of six agents solved the problem, while the other two could not learn the correct policy. The training curve in fig. 8.5 further showed that the training with the quantum computer needs more iterations than the simulator. However, due to the limited number of agents, the open question remains, whether the obtained accuracy on the quantum hardware is closer to 50% or to 83%, since $\frac{4}{6}$ is right in the middle.

The question was answered by increasing the experiments' granularity by using 12 instead of six agents, and the training was repeated 12 times instead of just one. In order to achieve this, the following steps have been performed by:

1. Obtaining the optimal hyper-parameters for the quantum REINFORCE algorithm with 12 agents on the 2D4 IG.
2. Showing that they are comparable to the results with only six agents and increased the results' granularity.

3. Proving that the evaluation of the trained agents on the quantum computer yields the same results.
4. Analyzing that 2,000 circuit measurements on quantum hardware are enough to get a good trade-off between run-time on the quantum hardware and reliable results.
5. Performing the training 12 times.

Step 4 was included to get a trade-off between accuracy and algorithms run-time since more than 20,000 quantum circuits need to be computed during the training, and different quantum computers were used for the training. Besides the fact that 2,000 circuits provide good accuracy, the analysis further revealed that the computational errors between the two quantum computers is lower compared to the simulator. This observation indicates that there might be differences in the optimal weights of the QVC between the simulator and real hardware. Therefore, better results could be obtained if the parameters are fine-tuned to the quantum hardware. However, as already argued, this is not possible at the moment.

The final step provided the required results to answer whether the training accuracy is better or worse than 66%. Averaging over the 12 training runs and comparing it to the simulator results in fig. 8.14 revealed that the accuracy is about 55%. Thus, seven out of 12 agents learned to solve the problem on the real quantum hardware on average. Further analysis revealed that some agents (like agent II) never solved the problem, while others constantly performed well, e.g., agent V and X. The exact reason for these differences could not be identified.

There can be various reasons why the training deviates from the one observed with a simulator. The most likely reason might be that the gradient computation does not take the quantum computer's noise into account, which leads to a sub-optimal solution since the optimization gets stuck in a local optimum. This is likely to be solved once the complete training is performed on the quantum computer. However, there are as well first results that indicate that noisy computations can complicate the learning process [Wang et al., 2020].

It might be possible that adjusting the agents' hyper-parameters could lead to better results since the learning rate determines the amount of change in the trainable weights. A lower learning rate could stabilize the training. However, performing hyper-parameter optimization in the spirit of [Bergstra et al., 2011] on the quantum computer would again lead to too many quantum circuit computations that are not possible with today's hardware resources.

This chapter aimed to answer the second research question:

Can such methods be trained on today's NISQ devices without a quantum simulator, and does quantum learning provide potential benefits, like improved convergence, compared to learning with a simulator?

It was shown that training on the real quantum hardware is possible. However, some simplifications are required, e.g., computing the gradients on a simulator. The results show that in half of the cases, the noisy quantum computer still obtains the optimal solution. However, the algorithm needs more time to converge to the optimal solution. Therefore, with today's NISQ computers, no beneficial behavior could be observed yet. However, it should be taken into account that the algorithms were optimized to the quantum simulators and not to the actual quantum hardware. As the intermediate analysis showed, there might be differences between the optimal parameter-setting between NISQ devices and simulators, which could improve the results. Furthermore, improved error correction algorithms and hardware are likely to improve the results in the future [Resch and Karpuzcu, 2019].

Finally, it should be considered that, despite the sub-optimal final results, since only about 55% of the agents learned to solve the problem, the results are, in the majority of cases, good enough to provide the right solution. These results could already be obtained even with the limited capacity of today's quantum computers. Although a simulator was necessary to compute the gradients for the training, the critical step that produces the training data was run on the quantum computer. If something like automatic differentiation (as in [Abadi et al., 2016]) were implemented on the quantum computer, no simulator would be needed, and the complete training could be performed on the quantum computer. This could be a reasonable step to increase the performance of the next generation of quantum hardware that is specialized for machine learning and optimization tasks.

Future Steps in Quantum Reinforcement Learning for Complex Scenarios

9

The last two chapters focused on answering the two research questions. The results show that *quantum REINFORCE trumps* the *classical methods* in terms of the number of training iterations until the optimal solution is obtained. It further revealed that the quantum method is memory efficient. This is in line with the results of [Chen et al., 2020] for the reduced memory usage in value-based RL (DQN) with a quantum approximator.

Performing the full training on a real quantum computer, as shown in section 8.2, is not possible at the moment. However, once the algorithm is trained, it will provide the same results when computed on the quantum hardware. Furthermore, it has been shown that most of the results are still correct if the episodes were generated on a quantum system while the training is performed with a simulator. However, several observations have been made on how quantum RL could be improved in the future during the experiments.

This chapter summarizes these observations to point out possible future steps to enhance the quantum REINFORCE method. It further shares some practical issues that arose while working with the IBM quantum hardware, which can be considered while developing other quantum policy gradient algorithms. The first section focuses on possible improvements that better quantum computers might bring in the future. The next section concentrates on the encoding function that is used to transfer the classical data onto the quantum computer [Schuld and Petruccione, 2018]. During this work, it has been shown that the data encoding significantly impacts the training results. Finally, the role of the QVC, which is the central technique in the quantum REINFORCE algorithm, is discussed.

9.1   Characteristics of NISQ devices

This work relies on IBM's quantum computers, which can be accessed through the *IBM Quantum Experience*[1]. The IBM Quantum Experience is a cloud-based platform to provide access to the full range of IBM's quantum hardware. It is further used to manage the accessible device, e.g., reserve time on a specific device. As previously mentioned, the Universität der Bundeswehr München is a partner in the IBM Quantum Network, which has the benefit of full access to all quantum systems. At the moment, 14 quantum computers and five powerful quantum simulator are available. The most extensive quantum systems have 65 qubits, and the smallest is a single-qubit system. Furthermore, there are quantum computers with 27, 7, and 5 qubits.

Quantum algorithms can be sent directly to the systems using the IBM quantum programming language Qiskit. The algorithm is sent to a specific system and stored in a queue until it is scheduled for execution. The algorithm's results are sent back after the computations have been carried out. This process is depicted in fig. 9.1. If there is a high demand on the systems, it can take up to several hours or even days until the submitted code is executed. This leads to long program run-times, which prevent a fair comparison of classical and quantum algorithms. Fortunately, some quantum systems can be reserved for up to 12 hours each month through the IBM quantum network. This helps to execute longer computations. However, it is not suitable for real applications yet.

IBM will tackle this problem in the near term future with an extension that will be called Qiskit-Runtime [IBM, 2021a]. With Qiskit-Runtime, it should be possible to send a whole program to the IBM Quantum Experience, and both quantum and classical computations will be performed on their systems. In theory, this should reduce the computational time since the waiting time in the job queue will be removed. The extent to which this will be helpful in the future still has to be shown. However, it points out that there are still many possibilities that can help to increase computational speed and maybe even reduce errors.

Another problem with the *cloud-based access* is an unstable connection to the cloud resources. If, for example, an algorithm has a long waiting time in the queue, the connection to IBM's web servers may get closed for some reason. This leads to an error in the training or evaluation program, and the whole computation needs to be repeated. If this happens near the end of the run-time, it is especially unpleasant. This problem can be countered with sufficient error handling in the program. However,

---

[1] https://quantum-computing.ibm.com/

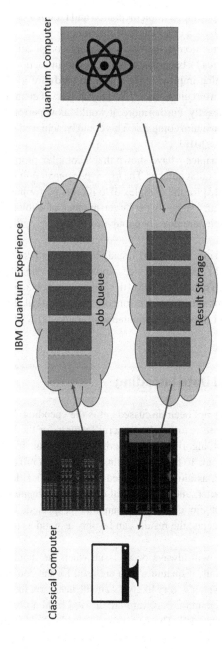

Figure 9.1 Visualizing the interaction with IBM's quantum computers

this complicates the algorithm's implementation and requires more resources on the classical computer to have a complete fault-tolerant program.

The last point to mention is the costly gradient computation on the quantum computer. This has already been discussed in length in section 8.2. Nevertheless, it should be stressed that improving the quantum hardware to faster compute the gradients for a quantum circuit will decrease the run-time of quantum machine learning algorithms significantly. Furthermore, it would allow performing the complete QVC training on the quantum computer. This could be achieved with a system-stack similar to [Abadi et al., 2016].

The results from chapter 7 have shown that a complex problem can already be solved with a moderate qubit number. The *hyper-parameter optimization* in the spirit of [Bergstra et al., 2013] for the complex 3D6 IG was performed with a maximum of 12 qubits. The results show that a larger number of qubits does not improve the training. Nine qubits are already enough to obtain the optimal solution in the complex 3D scenario.

Computations with nine qubits can already be performed on multiple IBM quantum computers, and the 65 qubits device should provide enough reserves to solve even larger problems. Thus, larger NISQ devices are unlikely to improve the results. However, reduced error rates through error correction [Preskill, 2018], more reliable and stable access, and improved computations (gradients) would be more helpful than larger systems.

9.2   Improved Data Encoding

Another point that has not been discussed yet is the encoding. While the quantum hardware's sole purpose is to compute useful features for the linear model in the quantum REINFORCE algorithm, the encoding scheme transfers the classical data onto the quantum system. To find an encoding function that reliably fulfills this task is of utter importance, as already discussed in section 6.1.1. It provides the data to the quantum computer, and if the initial data on the quantum processor is not well suited for the problem, good results can not be expected. This is similar to all machine learning concepts; the results can be only as good as the data provided to an algorithm.

The data encoding will always be an issue when the input is derived from a classical computer. Thus, if quantum data are used for the computations, this step can be omitted. However, it is very likely that in the near term future, most of the RL problems studied on quantum hardware are derived from a classical problem and therefore need to be encoded. The research of [Suzuki et al., 2020], on which the *encoding scheme* in this work is based, provides a rigorous analysis of five encoding functions on four different two-dimensional data sets. Although the results help to understand the effect of a specific data encoding better, it is of limited informative value since it is unknown if these results can be generalized to different data sets. Therefore, more research on the effect of the encoding function onto the quantum algorithm is needed and to solve the question of how to achieve an encoding function for a specific problem.

The encoding proposed by [Suzuki et al., 2020], on which the feature encoding scheme in this work is based, only analyzed the encoding for two-qubit systems. Thus, their results only hold in the 2D case with two features. For the 3D problem, it is not known whether the scheme actually provides a feasible encoding. Since the 3D6 Information Game results seem like the encoding generalizes to three qubits, it is not changed. However, it is not known yet if there might be a better encoding for the 3D case.

Furthermore, it would be beneficial to analyze the encoding scheme in the setting of RL. Unlike other machine learning disciplines, RL has other requirements on the data to draw reasonable conclusions from them. Therefore, it is helpful to not only rely on the results that have been made in the other two disciplines but to examine possible ways to encode the data in RL.

Suppose the quantum REINFORCE method or possibly other quantum RL methods should be used to solve problems with a visual input like the Atari games. In that case, special care needs to be taken to transfer the information contained in the images to the quantum computer. Some approaches study quantum algorithms similar to classical convolutional neural networks, which are state-of-the-art for image processing in machine learning. For example, the DQN algorithm [Mnih et al., 2015] that solves most of the Atari games utilizes several convolutional layers in the neural network to process the image obtained from the Atari game. Thus, to solve problems where the state is encoded as an image, different encoding functions will be needed, and the research into this direction is very scarce at the moment. It will be an exciting and, at the same time, vital field to drive the developments in quantum RL forwards.

The above discussion shows that the encoding is essential to all quantum RL algorithms. Multiple points can possibly provide improvements to the quantum method. The first point is to investigate other encoding functions that could be beneficial for the training of the QVC. The other direction is to search for other encoding methods, which transfer other forms of data like images onto the quantum computer. However, the results have shown that the used encoding scheme already achieved good results that can compete with the classical methods.

9.3   Analysis of Quantum Variational Circuits in Quantum Policy Gradient methods

The QVC is the central element of the quantum REINFORCE method. It is the method that allows the classical REINFORCE algorithm to be quantized. While the classical algorithm relies on a neural network or a linear model to approximate the policy, the QVC is a quantum hybrid approximator that can be used for the same task. However, there exist other methods that could be used for the approximation. For example, [Verdon et al., 2018] provides an overview of quantum neural networks, while a recent study focuses on achieving a quantum perceptron neural network [Sharma et al., 2020].

A quantum perceptron neural network could as well replace the QVC to build a different quantum policy gradient method. Today's research is in an early phase to explore different ways to use quantum computers. Therefore, it is likely that another quantum approximation algorithm will be found in the future that outperforms QVC. However, at this stage, there is a decent understanding of QVC, and efficient algorithms exist to train them. Variational quantum DQL [Chen et al., 2020], for example, has already shown that QVCs are a promising way that achieves good results in RL. This study's results strengthen these findings since in chapter 7 the quantum REINFORCE methods have proven to be memory efficient and to reduce the number of training iterations to solve the problem.

These results show that at the moment, QVC provides a promising path to achieve quantum RL that can solve complex problems. Furthermore, the quantum REIN-FORCE algorithm's concepts as the first known quantum policy gradient (as identified in chapter 4) method can be used to quantize other policy gradient methods. For example, based on this work, REINFORCE with a baseline could be directly implemented [Sutton and Barto, 2018]. Thus, although there might be numerous other quantum approximation methods, VQC provides a good starting point for quantum RL. Although it might be that improved QVC methods will be discovered in the next years, or a different method will be more effective. However, the quantum REINFORCE algorithm and other policy gradient methods can be easily adapted to better approximators based on this work.

Conclusion

10

The introduction in chapter 1 started by pointing out that conflicts and, more precisely, attacker-defender scenarios have been and still are constantly present on earth. A short review of the path from early conflicts on the predator-prey level to highly complicated situations like the theft of the Mona Lisa in 1911 or today's security systems and strategies at vulnerable infrastructure like airports is given. Furthermore, it was argued that classical techniques could hardly be applied to derive solutions for such scenarios. However, Reinforcement Learning (RL) was identified as a state-of-the-art theory to solve complex attacker-defender scenarios [Moll and Kunczik, 2019]. This was underpinned by recapturing its latest successes like a computer beating the world champion in the game GO [Silver et al., 2016, Silver et al., 2017], solving all 57 Atari 2600 games with a single algorithm [Badia et al., 2020] or competing with the top players in the computer-game StarCraft II [Vinyals et al., 2019].

As argued in the introductory chapter, the successes in RL come at a price of high computational cost. Enormous computational resources in the form of CPUs and GPUs are needed during the training process to obtain these results, and long run-times during the training are needed. A new emerging technology that could be able to overcome these drawbacks is quantum computing. With their computations based on the theory of quantum mechanics, and therefore relying on qubits (appendix A.1 in the electronic supplementary material) as the information-carrying unit [Scherer, 2019]. They can leverage on state superposition and entanglement. The former means that through the use of probability amplitudes, more information can be

---

Supplementary Information The online version contains supplementary material available at (https://doi.org/10.1007/978-3-658-37616-1_10).

L. Kunczik, *Reinforcement Learning with Hybrid Quantum Approximation in the NISQ Context*, https://doi.org/10.1007/978-3-658-37616-1_10

processed in a single computation, while the latter allows interactions between different qubits. Due to those possible advantages in quantum computing, it was identified as a possible solution to today's RL's computational requirements that constrain solving even more complex attacker-defender problems.

Chapter 2 introduced a new form of attacker-defender scenario called the Information Game that is studied throughout the work. It is motivated by a situation in which the attackers' goal is to extract information from a physical location. The defender uses a fixed strategy to protect the information against the attacker. The fixed defending strategy is similar to security measures on airports, where control points, cameras, and other security measures have to be planed during the construction of the building. The Information Game provides an advantage for analyzing and comparing different RL algorithms since it can be scaled easily to different problemsizes and settings. Within this work, it is extended from a simple 2D problem to a complex 3D problem with increased state and actions space.

The following chapter first reviewed the basic theory of RL and focused then on approximation methods in RL. Two standard techniques that utilize function approximation in RL, namely deep Q-networks (DQN) and policy gradient methods, were introduced. The former utilizes neural networks to approximate the action-value function for each state and action pair in the environment. The latter relies on function approximators like neural networks to directly approximate the policy. Although both methods rely on neural networks, both have significant advantages or disadvantages, which have been discussed in the context of attacker-defender scenarios in section 3.4. The discussion concluded that policy gradient methods can be advantageous in attacker-defender situations since the policy is not as complex to represent as the value function, which allows less involved approximators.

Based on the introduction of Reinforcement Learning, chapter 4 reviewed the literature on RL methods that utilize quantum computing. The literature review identified three major streams in quantum RL:

- Quantum Reinforcement Learning
- Projective Simulation
- Quantum Hybride Approximation methods.

Quantum Reinforcement Learning is, by publication date, the first occurrence of quantum computing in RL, although it does not utilize quantum hardware for computations. Its main idea is to base the action-representation and actions-selection on quantum states stored on classical hardware. Thus, Quantum Reinforcement Learning first introduced the idea of combining quantum computing and RL. However, it never evolved to a truly quantum computing algorithm.

Projective simulation is an RL framework that allows performing computations on quantum and classical hardware. However, today's research indicates that it is not practical to be used on quantum hardware at the moment, and it is not clear if this problem will be solved in the future [Sriarunothai et al., 2019]. Quantum hybrid approximation methods do not suffer from those implementation problems. They combine classical RL algorithms with quantum approximation methods like quantum variational circuits (QVC). The most successes have been accomplished by replacing the neural network in DQN with a VQC [Chen et al., 2020, Lockwood and Si, 2020].

This builds the foundation for identifying the two research questions that are answered with this work. QVCs have been successfully applied in DQN to solve different problems. However, as argued in section 3.4 policy gradient methods provide some benefits in attacker-defender scenarios. Therefore, the two research questions are:

- Can QVC be utilized to approximate the policy in policy gradient RL methods directly, and how do they perform compared to their classic counterparts?
- Can such methods be trained on today's NISQ devices without a quantum simulator, and does quantum learning provide potential benefits, like improved convergence, compared to learning with a simulator?

The first research question aims to extend the idea of [Chen et al., 2020] to policy gradient RL, which has not been done before. The second research question focuses on evaluating the quantum algorithm on real quantum hardware since today's quantum computers are still under development and do not perform perfect computations [Preskill, 2018]. It is, therefore, of special interest to evaluate if the new algorithm is robust against hardware errors, like in [Wang et al., 2020].

In order to answer the research questions, chapter 5 provided an in-depth introduction to QVC and its underlying theory. Based on this background, the new quantum REINFORCE algorithm was developed in chapter 6. This was done in three steps: The first step replaced the neural network with the QVC. The other two steps introduced further changes (conditional training and policy storage) to improve the algorithms' performance in attacker-defender scenarios. This derivation was concluded by a detailed discussion of the utilized QVC and some implementation details on the experimental setup.

Chapter 7 answered the first research question by comparing and analyzing the convergence and memory usage of the standard Q-learning section 3.2 as a baseline with the classical, and the quantum REINFORCE algorithm on the three Information Game instances. In order to achieve a fair comparison, the algorithms' hyper-

parameters were optimized in a first step to ensure optimal performance. To analyze
the convergence rate of the algorithms, the random behavior in the initial training phase was removed to measure the time (number of episodes) between the first
observed success and the convergence to the optimal solution. The analysis revealed
that the quantum REINFORCE algorithm needed the least iterations until it converged in all three problem instances. This is a surprising result since it showed
even better convergence behavior than Q-learning, which usually is very effective
in small problem settings.

In terms of memory consumption, the quantum algorithm outperformed the other
two algorithms again. In the small 2D4 setting, it needed half the trainable weights
than the classical algorithms. This advantage improved with growing problem size
to about 25% in the complex 3D6 instance compared to its classical counterpart
and less than 10% compared to Q-learning. Thus, the first research question was
answered successfully since it was possible to utilize QVC in policy gradient RL,
and it further showed improved convergence and reduced memory requirements.

The following chapter 8 answered the second research question, whether it is
possible or not to run those algorithms on real quantum hardware like IBM quantum
computers. This was done in multiple steps. First, the trained algorithm was evaluated on the quantum hardware to ensure that the quantum computer can reproduce
the simulators' results. This was successfully confirmed by first computing the optimal policy once and then multiple times. Next, the training was implemented on the
quantum hardware. In order to achieve this, the quantum REINFORCE algorithm
needed some adjustments since the gradient computation would be too costly on
the real hardware. Therefore, an additional policy to compute the gradients on a
simulator was introduced.

The training results for the 2D4 Information Game instance on the real quantum
computer achieved an accuracy of about 65%. Due to the low resolution, stemming
from the small number of agents, it raised the question of whether the results are
closer to 50% or 90%. Therefore, the number of agents was double to increase
the level of detail, and the optimal hyper-parameters had to be obtained again.
The previous experiment to ensure that the policy provides the same results when
computed on the quantum hardware was successfully repeated. For a more detailed
picture, the evaluated policies from different quantum devices and the simulator
were compared. This analysis showed that the results from the different quantum
computers show lower inter deviations compared to the simulator.

Observing the reduced errors between the two quantum systems indicates that
optimizing the algorithms hyper-parameters to the quantum computer could provide
better results. Thus, there seem to be differences between the optimal simulator and
the real devices that might be countered with full learning on the quantum systems.

However, due to the limited access and computational time, this question could not be addressed within this work and should be analyzed in future work.

Finally, the training on the quantum hardware with an increased number of agents was performed. The training was repeated 12 times to obtain more detailed results. The result as depicted in fig. 8.14 show that only an accuracy of about 55% could be achieved with the increased resolution. Further analysis revealed that some agents consistently performed well during the 12 training runs while others did not. The data could not explain this behavior. However, two ways to approach this problem in the future have been identified. First, it should be examined if error correction techniques could improve the results. The other approach would be to check if performing the complete training on the quantum system (including the gradient computation) will provide better results. This could potentially include a hyper-parameter search on the quantum hardware to increase the accuracy.

To conclude the second research question, it has been shown that some simplifications are needed to perform the training on the quantum hardware. The results further show that only about half the agents converged to the optimal solution, and it took more iterations to converge. Thus, today's hardware does not provide benefits in the learning process of the algorithm. If the full training would be possible, potentially with hyper-parameter search [Bergstra et al., 2011], this result could be changed. However, this is not yet possible with the technology available today, and this question is still open for future research.

The final chapter summarized practical observations that have been made during the experiments, limitations that have been obtained, and possible ways to improve the work in the future. They have been split into three areas: The first deals with IBM's quantum hardware and its access through the cloud. The main improvement that would help in quantum RL would be a more cost-effective way to compute gradients (similar to [Abadi et al., 2016]) and to reduce waiting time in the job queue before the computation will be performed. During this work, the maximally observed waiting time was more than a week, which slows down computations with several thousand jobs that need to be computed.

The second area that provides possibilities for improvement is the data encoding for the QVC. Special circuits are needed to encode the classical data into quantum states. The encoding scheme utilized in this work has shown to be successful in different tasks [Suzuki et al., 2020]. However, it is yet not fully understood how the QVC performance depends on the encoding, although some research shows that it is crucial for success [Schuld et al., 2021].

Finally, the QVC itself and its circuits layout is still an open question in research, and it might be beneficial to examine other quantum approximation methods like quantum perceptron neural network [Sharma et al., 2020]. However, QVC is a good choice at the moment since it is known that they have the universal approximation property [Goto et al., 2020]. Thus, there are different ways to improve the quantum REINFORCE method and implement other policy gradient methods.

To conclude, this work introduced quantum REINFORCE as the first known policy gradient RL algorithm in quantum computing, as discussed in chapter 4. It was further shown that the quantum method improves convergence and reduces memory consumption compared to its classical counterpart or the value-based Q-learning algorithm. Thus, utilizing a QVC in policy gradient RL, like REINFORCE, can solve more complex problems with fewer hardware requirements than are needed classically. Furthermore, this work contributed to the first known approach to performing quantum RL training on real quantum hardware, compared to only evaluating the trained algorithm as in [Chen et al., 2020].

These experiments were possible by adapting the algorithm to IBM's quantum hardware. Despite the considerable noise in today's quantum systems, the majority of experiments obtained the optimal solution. When error mitigation techniques improve the quality of the computations in the future, quantum RL will be a prosperous alternative to classical computers. This work has shown that employing quantum approximation techniques in RL improves the performance while reducing the hardware requirements in terms of memory usage at the same time. Thus, quantum RL provides a fruitful path to solve even more challenging problems in the context of complex attacker-defender scenarios.

Bibliography

[Abadi et al., 2016] Abadi, M., Agarwal, A., Barham, P., Brevdo, E., Chen, Z., Citro, C., Corrado, G. S., Davis, A., Dean, J., Devin, M., et al. (2016). TensorFlow: Large-Scale Machine Learning on Heterogeneous Distributed Systems. 1603.04467.

[Abbas et al., 2021] Abbas, A., Sutter, D., Zoufal, C., Lucchi, A., Figalli, A., and Woerner, S. (2021). The power of quantum neural networks. *Nature Computational Science*, 1(6):403–409, 2011.00027.

[Abraham et al., 2019] Abraham, H., AduOffei, Agarwal, R., Akhalwaya, I. Y., Aleksandrowicz, G., Alexander, T., Amy, M., Arbel, E., et al. (2019). Qiskit: An Open-source Framework for Quantum Computing.

[Adachi and Henderson, 2015] Adachi, S. H. and Henderson, M. P. (2015). Application of Quantum Annealing to Training of Deep Neural Networks. *arXiv*, 1510.06356.

[Akiba et al., 2019] Akiba, T., Sano, S., Yanase, T., Ohta, T., and Koyama, M. (2019). Optuna: A Next-generation Hyperparameter Optimization Framework. *Proceedings of the ACM SIGKDD International Conference on Knowledge Discovery and Data Mining*, pages 2623–2631, 1907.10902.

[Alderson et al., 2011] Alderson, D. L., Brown, G. G., Carlyle, W. M., and Wood, R. K. (2011). Solving Defender-Attacker-Defender Models for Infrastructure Defense. In *12th INFORMS Computing Society Conference*, pages 28–49. INFORMS.

[Arora and Barak, 2009] Arora, S. and Barak, B. (2009). *Computational Complexity: A Modern Approach*. Cambridge University Press, USA, 1st edition.

[Arute et al., 2019] Arute, F., Arya, K., Babbush, R., Bacon, D., Bardin, J. C., Barends, R., Biswas, R., Boixo, S., Brandao, F. G., Buell, D. A., et al. (2019). Quantum supremacy using a programmable superconducting processor. *Nature*, 574(7779):505–510.

[Ashton, 1997] Ashton, T. S. (1997). *The Industrial Revolution 1760–1830*. Number 9780192892898 in OUP Catalogue. Oxford University Press.

[Badia et al., 2020] Badia, A. P., Piot, B., Kapturowski, S., Sprechmann, P., Vitvitskyi, A., Guo, D., and Blundell, C. (2020). Agent57: Outperforming the Atari Human Benchmark. *arXiv*, 2003.13350.

[Baker et al., 2019] Baker, B., Kanitscheider, I., Markov, T., Wu, Y., Powell, G., McGrew, B., and Mordatch, I. (2019). Emergent Tool Use From Multi-Agent Autocurricula. *arXiv*, 1909.07528.

[Barto et al., 1983] Barto, A. G., Sutton, R. S., and Anderson, C. W. (1983). Neuronlike Adaptive Elements That Can Solve Difficult Learning Control Problems. *IEEE Transactions on Systems, Man and Cybernetics*, pages 834–846.

[Beer et al., 2020] Beer, K., Bondarenko, D., Farrelly, T., Osborne, T. J., Salzmann, R., Scheiermann, D., and Wolf, R. (2020). Training deep quantum neural networks. *Nature Communications 2020 11:1*, 11(1):1–6.

[Bellemare et al., 2013] Bellemare, M. G., Naddaf, Y., Veness, J., and Bowling, M. (2013). The Arcade Learning Environment: An evaluation platform for general agents. *Journal of Artificial Intelligence Research*, pages 253–279, 1207.4708.

[Bellman, 1957] Bellman, R. (1957). *Dynamic Programming*. Princeton University Press, Princeton.

[Bengtson, 2002] Bengtson, S. (2002). Origins and Early Evolution of Predation. *The Paleontological Society Papers*, 8:289–318.

[Benioff, 1980] Benioff, P. (1980). The computer as a physical system: A microscopic quantum mechanical Hamiltonian model of computers as represented by Turing machines. *Journal of Statistical Physics*, 22(5):563–591.

[Bergholm et al., 2018] Bergholm, V., Izaac, J., Schuld, M., Gogolin, C., Alam, M. S., Ahmed, S., Arrazola, J. M., Blank, C., Delgado, A., Jahangiri, S., McKiernan, K., Meyer, J. J., Niu, Z., Száva, A., and Killoran, N. (2018). PennyLane: Automatic differentiation of hybrid quantum-classical computations. *arXiv*, 1811.04968.

[Bergstra et al., 2011] Bergstra, J., Bardenet, R., Bengio, Y., and Kégl, B. (2011). Algorithms for Hyper-Parameter Optimization. In Weinberger, J. S.-T., Zemel, R., Bartlett, P., Pereira, F., and Q., K., editors, *Advances in Neural Information Processing Systems*. Curran Associates, Inc.

[Bergstra et al., 2013] Bergstra, J., Yamins, D., and Cox, D. (2013). Making a science of model search: Hyperparameter optimization in hundreds of dimensions for vision architectures. In Dasgupta, S. and McAllester, D., editors, *Proceedings of the 30th International Conference on Machine Learning*, volume 28 of *Proceedings of Machine Learning Research*, pages 115–123, Atlanta, Georgia, USA. PMLR.

[Biamonte et al., 2017] Biamonte, J., Wittek, P., Pancotti, N., Rebentrost, P., Wiebe, N., and Lloyd, S. (2017). Quantum machine learning. *Nature*, 549(7671):195–202, 1611.09347.

[Briegel and De Las Cuevas, 2012] Briegel, H. J. and De Las Cuevas, G. (2012). Projective simulation for artificial intelligence. *Scientific Reports*, 2(1):400, 1104.3787.

[Brockman et al., 2016] Brockman, G., Cheung, V., Pettersson, L., Schneider, J., Schulman, J., Tang, J., and Zaremba, W. (2016). Openai gym. arXiv:1606.01540.

[Brown et al., 2006] Brown, G., Carlyle, M., Salmerón, J., and Wood, K. (2006). Defending critical infrastructure. *Interfaces*, 36(6):530–544.

[Cao, 2007] Cao, X.-R. (2007). Markov Decision Processes. In *Stochastic Learning and Optimization*, pages 183–252. Springer US, Boston, MA.

[Cao et al., 2019] Cao, Y., Romero, J., Olson, J. P., Degroote, M., Johnson, P. D., Kieferová, M., Kivlichan, I. D., Menke, T., Peropadre, B., Sawaya, N. P. D., Sim, S., Veis, L., and Aspuru-Guzik, A. (2019). Quantum Chemistry in the Age of Quantum Computing. *Chemical Reviews*, 119(19):10856–10915.

[Chen and Dong, 2012] Chen, C. and Dong, D. (2012). Quantum parallelization of hierarchical Q-learning for global navigation of mobile robots. In *Proceedings of 2012 9th IEEE International Conference on Networking, Sensing and Control*, pages 163–168. IEEE.

[Chen et al., 2008] Chen, C., Yang, P., Zhou, X., and Dong, D. (2008). A Quantum-inspired Q-learning Algorithm for Indoor Robot Navigation. In *2008 IEEE International Conference on Networking, Sensing and Control*, pages 1599–1603. IEEE.

[Chen et al., 2006] Chen, C. L., Dong, D. Y., and Chen, Z. H. (2006). Quantum Computation for action selection using Reinforcement Learning. *International Journal of Quantum Information*, 04(06):1071–1083.

[Chen et al., 2020] Chen, S. Y.-C., Yang, C.-H. H., Qi, J., Chen, P.-Y., Ma, X., and Goan, H.-S. (2020). Variational Quantum Circuits for Deep Reinforcement Learning. *IEEE Access*, 8:141007–141024, 1907.00397.

[Cheng et al., 2020] Cheng, H.-P., Deumens, E., Freericks, J. K., Li, C., and Sanders, B. A. (2020). Application of Quantum Computing to Biochemical Systems: A Look to the Future. *Frontiers in Chemistry*, 0:1066.

[Clausen and Briegel, 2018] Clausen, J. and Briegel, H. J. (2018). Quantum machine learning with glow for episodic tasks and decision games. *Physical Review A*, 97(2):022303.

[Copeland, 2004] Copeland, B. J. (2004). *The Essential Turing*. Oxford University Press.

[Crawford et al., 2018] Crawford, D., Levit, A., Ghadermarzy, N., Oberoi, J. S., and Ronagh, P. (2018). Reinforcement Learning Using Quantum Boltzmann Machines. *Quantum Information and Computation*, 18(1–2):51–74.

[Cross et al., 2019] Cross, A. W., Bishop, L. S., Sheldon, S., Nation, P. D., and Gambetta, J. M. (2019). Validating quantum computers using randomized model circuits. *Physical Review A*, 100(3):032328, 1811.12926.

[Crosson and Harrow, 2016] Crosson, E. and Harrow, A. W. (2016). Simulated Quantum Annealing Can Be Exponentially Faster Than Classical Simulated Annealing. In *2016 IEEE 57th Annual Symposium on Foundations of Computer Science (FOCS)*, pages 714–723. IEEE, 1601.03030.

[Daoyi Dong et al., 2008] Daoyi Dong, Chunlin Chen, Hanxiong Li, and Tzyh-Jong Tarn (2008). Quantum Reinforcement Learning. *IEEE Transactions on Systems, Man, and Cybernetics, Part B (Cybernetics)*, 38(5):1207–1220, 0810.3828.

[Darwin, 1859] Darwin, C. (1859). *On the origin of species by means of natural selection, or, The preservation of favoured races in the struggle for life*. J. Murray, London.

[Denil and De Freitas, 2011] Denil, M. and De Freitas, N. (2011). Toward the Implementation of a Quantum RBM. In *NIPS Deep Learning and Unsupervised Feature Learning Workshop, vol. 5*.

[Diuk et al., 2006] Diuk, C., Strehl, A. L., and Littman, M. L. (2006). A hierarchical approach to efficient reinforcement learning in deterministic domains. In *Proceedings of the Fifth International Joint Conference on Autonomous Agents and Multiagent Systems*, AAMAS '06, pages 313–319, New York, NY, USA. Association for Computing Machinery.

[Dong et al., 2005a] Dong, D., Chen, C., and Chen, Z. (2005a). Quantum Reinforcement Learning. In *Lecture Notes in Computer Science*, volume 3611, pages 686–689. Springer, Berlin, Heidelberg.

[Dong et al., 2006] Dong, D., Chen, C., and Li, H. (2006). Reinforcement Strategy Using Quantum Amplitude Amplification for Robot Learning. In *2007 Chinese Control Conference*, pages 571–575. IEEE.

[Dong et al., 2005b] Dong, D., Chen, C., Zhang, C., and Chen, Z. (2005b). An Autonomous Mobile Robot Based on Quantum Algorithm. In *Lecture Notes in Computer Science (includ-*

ing subseries Lecture Notes in Artificial Intelligence and Lecture Notes in Bioinformatics), volume 3801 LNAI, pages 393–398. Springer, Berlin, Heidelberg.

[Du et al., 2020] Du, Y., Hsieh, M.-H., Liu, T., and Tao, D. (2020). Expressive power of parametrized quantum circuits. *Physical Review Research*, 2(3):033125, 1810.11922.

[Dunjko et al., 2015] Dunjko, V., Friis, N., and Briegel, H. J. (2015). Quantum-enhanced deliberation of learning agents using trapped ions. *New Journal of Physics*, 17(2):023006, 1407.2830.

[Einstein et al., 1935] Einstein, A., Podolsky, B., and Rosen, N. (1935). Can quantum-mechanical description of physical reality be considered complete? *Physical Review*, 47(10):777–780.

[Fakhari et al., 2013] Fakhari, P., Rajagopal, K., Balakrishnan, S. N., and Busemeyer, J. R. (2013). Quantum inspired Reinforcement Learning in changing environment. *New Mathematics and Natural Computation*, 09(03):273–294.

[Farhi et al., 2014] Farhi, E., Goldstone, J., and Gutmann, S. (2014). A Quantum Approximate Optimization Algorithm. *arXiv*, 1411.4028.

[Farhi and Neven, 2018] Farhi, E. and Neven, H. (2018). Classification with Quantum Neural Networks on Near Term Processors. *arXiv*, 1802.06002.

[Feynman, 1982] Feynman, R. P. (1982). Simulating physics with computers. *International Journal of Theoretical Physics*, 21:467–488.

[Gao et al., 2017] Gao, X., Zhang, Z., and Duan, L. (2017). An efficient quantum algorithm for generative machine learning. *arXiv*, 1711.02038.

[Goto et al., 2020] Goto, T., Tran, Q. H., and Nakajima, K. (2020). Universal Approximation Property of Quantum Feature Map. *arXiv*, 2009.00298.

[Grover, 1996] Grover, L. K. (1996). A fast quantum mechanical algorithm for database search. In *Proceedings of the Twenty-Eighth Annual ACM Symposium on Theory of Computing*, STOC '96, pages 212–219. Association for Computing Machinery.

[Harrow et al., 2009] Harrow, A. W., Hassidim, A., and Lloyd, S. (2009). Quantum Algorithm for Linear Systems of Equations. *Physical Review Letters*, 103(15):150502.

[Hilbert, 2020] Hilbert, M. (2020). Digital technology and social change: the digital transformation of society from a historical perspective. *Dialogues in Clinical Neuroscience*, 22(2):189–194.

[Hoobler and Hoobler, 2009] Hoobler, T. and Hoobler, D. (2009). *The Crimes of Paris*. Little, Brown and Company.

[Hu and Wellman, 2003] Hu, J. and Wellman, M. P. (2003). Nash q-learning for general-sum stochastic games. *Journal of Machine Learning Research*, 4:1039–1069.

[IBM, 2021a] IBM (2021a). IBM's roadmap for building an open quantum software ecosystem | IBM Research Blog. https://www.ibm.com/blogs/research/2021/02/quantum-development-roadmap/. Last accessed on 2021-07-20.

[IBM, 2021b] IBM (2021b). On "quantum supremacy" | ibm research blog. https://www.ibm.com/blogs/research/2019/10/on-quantum-supremacy/. Last accessed on 2021-08-06.

[IBM, 2021c] IBM (2021c). System properties | IBM Quantum. https://quantum-computing.ibm.com/lab/docs/iql/manage/systems/properties. Last accessed on 2021-07-08.

[IEEE Std 754-2019, 2019] IEEE Std 754-2019 (2019). IEEE Standard for Floating-Point Arithmetic. *IEEE Std 754-2019 (Revision of IEEE 754-2008)*, pages 1–84.

[Kakade, 2003] Kakade, S. M. (2003). *On the Sample Complexity of Reinforcement Learning*. PhD thesis, University College London.

[Kandala et al., 2017] Kandala, A., Mezzacapo, A., Temme, K., Takita, M., Brink, M., Chow, J. M., and Gambetta, J. M. (2017). Hardware-efficient variational quantum eigensolver for small molecules and quantum magnets. *Nature 2017 549:7671*, 549(7671):242–246.

[Kearns, 1990] Kearns, M. J. (1990). *The Computational Complexity of Machine Learning*. PhD thesis, Harvard University, USA.

[Keeley, 1996] Keeley, L. H. (1996). *War Before Civilization*. Oxford paperbacks. Oxford University Press, USA.

[Kingma and Ba, 2015] Kingma, D. P. and Ba, J. L. (2015). Adam: A method for stochastic optimization. In *3rd International Conference on Learning Representations, ICLR 2015 – Conference Track Proceedings*. 1412.6980.

[Levit et al., 2017] Levit, A., Crawford, D., Ghadermarzy, N., Oberoi, J. S., Zahedinejad, E., and Ronagh, P. (2017). Free energy-based reinforcement learning using a quantum processor. *arXiv*, 1706.00074.

[Lockwood and Si, 2020] Lockwood, O. and Si, M. (2020). Reinforcement Learning with Quantum Variational Circuits. In *Proceedings of the AAAI Conference on Artificial Intelligence and Interactive Digital Entertainment*, volume 16, pages 245–251. AIIDE-20.

[Mari et al., 2020] Mari, A., Bromley, T. R., Izaac, J., Schuld, M., and Killoran, N. (2020). Transfer learning in hybrid classical-quantum neural networks. *Quantum*, 4:340, 1912.08278.

[Matignon et al., 2006] Matignon, L., Laurent, G. J., and Le Fort-Piat, N. (2006). Reward function and initial values: Better choices for accelerated goal-directed reinforcement learning. In Kollias, S. D., Stafylopatis, A., Duch, W., and Oja, E., editors, *Artificial Neural Networks – ICANN 2006*, volume 4131 LNCS, pages 840–849. Springer Verlag.

[Meng et al., 2006] Meng, X., Yuzhen, P., Quande, Y., and Ying, P. (2006). A Study of Multiagent Reinforcement Learning based on Quantum Theory. In *The Proceedings of the Multiconference on "Computational Engineering in Systems Applications"*, pages 1990–1993. IEEE.

[Metropolis and Ulam, 1949] Metropolis, N. and Ulam, S. (1949). The monte carlo method. *Journal of the American Statistical Association*, 44(247):335–341.

[Michie and Chambers, 1968] Michie, D. and Chambers, R. A. (1968). BOXES: An Experiment in Adaptive Control. In Dale, E. and Michie, D., editors, *Machine Intelligence*. Oliver and Boyd.

[Minsky, 1954] Minsky, M. L. (1954). *Theory of neural-analog reinforcement systems and its application to the brain-model problem*. PhD thesis, Princeton University.

[Mitarai et al., 2018] Mitarai, K., Negoro, M., Kitagawa, M., and Fujii, K. (2018). Quantum circuit learning. *Physical Review A*, 98(3):32309, 1803.00745.

[Mnih et al., 2015] Mnih, V., Kavukcuoglu, K., Silver, D., Rusu, A. A., Veness, J., Bellemare, M. G., Graves, A., Riedmiller, M., Fidjeland, A. K., Ostrovski, G., Petersen, S., Beattie, C., Sadik, A., Antonoglou, I., King, H., Kumaran, D., Wierstra, D., Legg, S., and Hassabis, D. (2015). Human-level control through deep reinforcement learning. *Nature*, 518(7540):529–533.

[Moll and Kunczik, 2019] Moll, M. and Kunczik, L. (2019). Two Perspectives on Playing Games: Reinforcement Learning vs Game Theory. In *Proceedings on the International Conference on Artificial Intelligence (ICAI)*, pages 60–61. The Steering Committee of The World Congress in Computer Science, Computer Engineering and Applied Computing (WorldComp).

[Moll and Kunczik, 2021] Moll, M. and Kunczik, L. (2021). Comparing quantum hybrid reinforcement learning to classical methods. *Human-Intelligent Systems Integration*, 3(1):15–
23.

[Neumann et al., 2020] Neumann, N. M., de Heer, P. B., Chiscop, I., and Phillipson, F. (2020).
Multi-agent reinforcement learning using simulated quantum annealing. In *Lecture Notes in
Computer Science (including subseries Lecture Notes in Artificial Intelligence and Lecture
Notes in Bioinformatics)*, volume 12142 LNCS, pages 562–575. Springer.

[Nielsen and Chuang, 2011] Nielsen, M. A. and Chuang, I. L. (2011). *Quantum Computation
and Quantum Information: 10th Anniversary Edition*. Cambridge University Press, USA,
10th edition.

[Niu et al., 2019] Niu, M. Y., Boixo, S., Smelyanskiy, V. N., and Neven, H. (2019). Universal
quantum control through deep reinforcement learning. *npj Quantum Information*, 5(1):1–8.

[Oguni et al., 2014] Oguni, K., Narisawa, K., and Shinohara, A. (2014). Reducing sample
complexity in reinforcement learning by transferring transition and reward probabilities.
*ICAART 2014 – Proceedings of the 6th International Conference on Agents and Artificial
Intelligence*, 1:632–638.

[Paparo et al., 2014] Paparo, G. D., Dunjko, V., Makmal, A., Martin-Delgado, M. A., and
Briegel, H. J. (2014). Quantum speedup for active learning agents. *Physical Review X*,
4(3):031002, 1401.4997.

[Pavlov, 1927] Pavlov, I. P. (1927). *Conditioned reflexes: an investigation of the physiological
activity of the cerebral cortex*. Oxford Univ. Press.

[Pemantle, 2007] Pemantle, R. (2007). A survey of random processes with reinforcement.
Probability Surveys, 4(1):1–79.

[Porotti et al., 2019] Porotti, R., Tamascelli, D., Restelli, M., and Prati, E. (2019). Coherent
transport of quantum states by deep reinforcement learning. *Communications Physics*,
2(1):1–9.

[Powell, 2007] Powell, W. B. (2007). *Approximate Dynamic Programming: Solving the Curses
of Dimensionality (Wiley Series in Probability and Statistics)*. Wiley-Interscience, USA.

[Preskill, 2012] Preskill, J. (2012). Quantum computing and the entanglement frontier. *arXiv*,
1203.5813.

[Preskill, 2018] Preskill, J. (2018). Quantum Computing in the NISQ era and beyond. *Quantum*, 2:79, 1801.00862.

[Resch and Karpuzcu, 2019] Resch, S. and Karpuzcu, U. R. (2019). Quantum Computing:
An Overview Across the System Stack. *arXiv*, 1905.07240.

[Rieffel and Polak, 2011] Rieffel, E. and Polak, W. (2011). *Quantum Computing: A Gentle
Introduction*. The MIT Press, 1st edition.

[Roffe, 2019] Roffe, J. (2019). Quantum Error Correction: An Introductory Guide. *Contemporary Physics*, 60(3):226–245, 1907.11157.

[Sallans and Hinton, 2004] Sallans, B. and Hinton, G. E. (2004). Reinforcement Learning
with Factored States and Actions. *Journal of Machine Learning Research*, 5:1063–1088.

[Sargent, 2000] Sargent, R. W. (2000). Optimal control. *Journal of Computational and Applied
Mathematics*, 124(1–2):361–371.

[Scherer, 2019] Scherer, W. (2019). *Mathematics of Quantum Computing*. Springer International Publishing, Cham.

[Schuld et al., 2020] Schuld, M., Bocharov, A., Svore, K. M., and Wiebe, N. (2020). Circuit-
centric quantum classifiers. *Physical Review A*, 101(3):032308, 1804.00633.

[Schuld and Killoran, 2019] Schuld, M. and Killoran, N. (2019). Quantum Machine Learning in Feature Hilbert Spaces. *Physical Review Letters*, 122(4):040504, 1803.07128.

[Schuld and Petruccione, 2018] Schuld, M. and Petruccione, F. (2018). *Supervised Learning with Quantum Computers*. Quantum Science and Technology. Springer International Publishing, Cham, 1203.5813.

[Schuld et al., 2015] Schuld, M., Sinayskiy, I., and Petruccione, F. (2015). An introduction to quantum machine learning. *Contemporary Physics*, 56(2):172–185, 1409.3097v1.

[Schuld et al., 2021] Schuld, M., Sweke, R., and Meyer, J. J. (2021). Effect of data encoding on the expressive power of variational quantum-machine-learning models. *Physical Review A*, 103(3):032430, 2008.08605.

[Scull, 1992] Scull, C. (1992). Lotte Hedeager. Iron Age societies: from tribe to state in northern Europe, 500 BC to AD 700. (Social Archaeology). Translated by John Hines. ix 274 pages, 95 figures. 1992. Oxford: Basil Blackwell: ISBN 0-631-17106-1 hardback £30. *Antiquity*, 66(253):989–990.

[Sharma et al., 2020] Sharma, K., Cerezo, M., Cincio, L., and Coles, P. J. (2020). Trainability of Dissipative Perceptron-Based Quantum Neural Networks. *arXiv*, 2005.12458.

[Silver et al., 2016] Silver, D., Huang, A., Maddison, C. J., Guez, A., Sifre, L., Van Den Driessche, G., Schrittwieser, J., Antonoglou, I., Panneershelvam, V., Lanctot, M., et al. (2016). Mastering the game of Go with deep neural networks and tree search. *Nature*, 529:484–489.

[Silver et al., 2017] Silver, D., Schrittwieser, J., Simonyan, K., Antonoglou, I., Huang, A., Guez, A., Hubert, T., Baker, L., Lai, M., Bolton, A., Chen, Y., Lillicrap, T., Hui, F., Sifre, L., Van Den Driessche, G., Graepel, T., and Hassabis, D. (2017). Mastering the game of Go without human knowledge. *Nature*, 550(7676):354–359.

[Sim et al., 2019] Sim, S., Johnson, P. D., and Aspuru-Guzik, A. (2019). Expressibility and entangling capability of parameterized quantum circuits for hybrid quantum-classical algorithms. *Advanced Quantum Technologies*, 2(12):1900070, 1905.10876.

[Sriarunothai et al., 2019] Sriarunothai, T., Wölk, S., Giri, G. S., Friis, N., Dunjko, V., Briegel, H. J., and Wunderlich, C. (2019). Speeding-up the decision making of a learning agent using an ion trap quantum processor. *Quantum Science and Technology*, 4(1):015014, 1709.01366.

[Statista, 2021] Statista (2021). Biggest companies in the world by market cap 2021. https://www.statista.com/statistics/263264/top-companies-in-the-world-by-market-capitalization/. Last accessed on 2021-06-21.

[Sung et al., 2020] Sung, K. J., Yao, J., Harrigan, M. P., Rubin, N. C., Jiang, Z., Lin, L., Babbush, R., and McClean, J. R. (2020). Using models to improve optimizers for variational quantum algorithms. *Quantum Science and Technology*, 5(4):044008, 2005.11011.

[Sutton and Barto, 2018] Sutton, R. S. and Barto, A. G. (2018). *Reinforcement Learning: An Introduction*. A Bradford Book, second edition.

[Suzuki et al., 2020] Suzuki, Y., Yano, H., Gao, Q., Uno, S., Tanaka, T., Akiyama, M., and Yamamoto, N. (2020). Analysis and synthesis of feature map for kernel-based quantum classifier. *Quantum Machine Intelligence*, 2(1):9, 1906.10467.

[Sweke et al., 2020] Sweke, R., Wilde, F., Meyer, J., Schuld, M., Faehrmann, P. K., Meynard-Piganeau, B., and Eisert, J. (2020). Stochastic gradient descent for hybrid quantum-classical optimization. *Quantum*, 4:314, 1910.01155.

[Szepesvári, 1997] Szepesvári, C. (1997). The Asymptotic Convergence-Rate of Q-Learning. In *Proceedings of the 10th International Conference on Neural Information Processing Systems*, NIPS'97, pages 1064–1070, Cambridge, MA, USA. MIT Press.

[Tan et al., 2009] Tan, J., Meng, X.-P., Wang, T., and Wang, S.-B. (2009). Multi-agent reinforcement learning based on quantum andant colony algorithm theory. In *2009 International Conference on Machine Learning and Cybernetics*, volume 3, pages 1759–1764. IEEE.

[Tesauro, 1995] Tesauro, G. (1995). Temporal difference learning and TD-Gammon. *Communications of the ACM*, 38(3):58–68.

[Thorndike, 1911] Thorndike, E. L. (1911). *Animal intelligence; experimental studies,.* The Macmillan Company,.

[Tsitsiklis and Van Roy, 1997] Tsitsiklis, J. N. and Van Roy, B. (1997). Analysis of temporal-difference learning with function approximation. In *Advances in Neural Information Processing Systems*, volume 42, pages 674–690.

[van Hasselt et al., 2015] van Hasselt, H., Guez, A., and Silver, D. (2015). Deep Reinforcement Learning with Double Q-learning. *30th AAAI Conference on Artificial Intelligence, AAAI 2016*, pages 2094–2100, 1509.06461.

[Verdon et al., 2018] Verdon, G., Pye, J., and Broughton, M. (2018). A Universal Training Algorithm for Quantum Deep Learning. *arXiv*, 1806.09729.

[Vinyals et al., 2019] Vinyals, O., Babuschkin, I., Czarnecki, W. M., Mathieu, M., Dudzik, A., Chung, J., Choi, D. H., Powell, R., Ewalds, T., Georgiev, P., et al. (2019). Grandmaster level in StarCraft II using multi-agent reinforcement learning. *Nature 2019 575:7782*, 575(7782):350–354.

[Wang et al., 2020] Wang, S., Fontana, E., Cerezo, M., Sharma, K., Sone, A., Cincio, L., and Coles, P. J. (2020). Noise-Induced Barren Plateaus in Variational Quantum Algorithms. *arXiv*, 2007.14384.

[Wang et al., 2019] Wang, T., Bao, X., Clavera, I., Hoang, J., Wen, Y., Langlois, E., Zhang, S., Zhang, G., Abbeel, P., and Ba, J. (2019). Benchmarking Model-Based Reinforcement Learning. *arXiv*, 1907.02057.

[Watkins and Dayan, 1992] Watkins, C. J. C. H. and Dayan, P. (1992). Q-learning. *Machine Learning*.

[Wichert, 2013] Wichert, A. (2013). *Principles of quantum artificial intelligence.* World Scientific Publishing Co.

[Wiering and van Otterlo, 2012] Wiering, M. and van Otterlo, M. (2012). *Reinforcement Learning.* Adaptation, Learning, and Optimization. Springer, Berlin, Heidelberg.

[Wikipedia, 2021] Wikipedia (2021). List of quantum processors. https://en.wikipedia.org/wiki/List_of_quantum_processors. Last accessed on 2021-07-27.

[Williams, 1992] Williams, R. J. (1992). Simple statistical gradient-following algorithms for connectionist reinforcement learning. *Machine Learning*, 8(3–4):229–256.

[Wittek, 2014] Wittek, P. (2014). *Quantum Machine Learning: What Quantum Computing Means to Data Mining.* Elsevier Inc.

[Xu et al., 2019] Xu, H., Li, J., Liu, L., Wang, Y., Yuan, H., and Wang, X. (2019). Generalizable control for quantum parameter estimation through reinforcement learning. *npj Quantum Information*, 5(1):1–8.

[Zi et al., 2009] Zi, S., Waldron, A., and Mair, V. H. (2009). *The Art of War: Sun Zi's Military Methods.* Columbia University Press.

Printed in the United States
by Baker & Taylor Publisher Services